Lecture Notes in Computer Science

Edited by G. Goos and J. Hartmanis

228

Applied Algebra, Algorithmics and Error-Correcting Codes

2nd International Conference, AAECC-2
Toulouse, France, October 1–5, 1984
Proceedings

Edited by Alain Poli

Springer-Verlag
Berlin Heidelberg New York London Paris Tokyo

Editor

Alain Poli
AAECC/LSI Lab., Université P. Sabatier
118, route de Narbonne, 31062 Toulouse Cédex, France

CR Subject Classifications (1985): B.4.5, G.2.0

ISBN 3-540-16767-6 Springer-Verlag Berlin Heidelberg New York
ISBN 0-387-16767-6 Springer-Verlag New York Berlin Heidelberg

Printed in Germany

Printing and binding: Beltz Offsetdruck, Hemsbach/Bergstr.
2145/3140-543210

PREFACE

The International Colloquium on Applied Algebra and Error Correcting Codes was born in Toulouse (France) in June 1983.

The acts of AAECC-1 are published in Discrete Mathematics (vol 56 n°2-3, Oct.85). The acts of AAECC-2 are contained in this volume.

From 48 talks, we have selected 23 accepted papers, after a (time consuming) system of multiple reviews. I thank those referees who agreed to contribute to the obtained result.

I also thank :

- Mr. A. Oisel and CII-HBull for their financial support,
- Mr. M. Combarnous Scientific Director of CNRS*, for CNRS's financial support,
- Mr. A. Dargent, Director of CNES** Informatic Center, for allowing us the use of the computers before and during the conference,
- The LSI laboratory and University P. Sabatier for their financial support.

As one knows, digitalized data are becoming increasingly important, particularly for transmissions.
For satellite transmissions, the CCSDS (Consultative Committee for Data Space System) had proposed a coding system for international transmissions (see : final report of contract AAECC/CNES n° 84/5417, 1985 (210 pages)).
Also, the target of RACE project is to define and realize a Broadband-IBC european network with security/privacy (cryptography) and reliability (error-correcting codes). AAECC lab. is a participant for the definition phase (in group n°2015).

As digitalized data are being more and more used for images/speech/files transmissions, theoretical tools and practical developments are necessary (for finite algebraic structures and for complexity analyses).
In particular, decomposition of algebras is an interesting topic because it is used for problems involved with complexity (see J. Heintz/J. Morgenstern), for constructive results on idempotents, multivariate codes (see : A. Poli, H. Imai, A. Poli/ C. Rigoni), for DFT's problems (see : T. Beth). Many other particular aspects of re-

---

* CNRS Centre National de la Recherche Scientifique
** CNES Centre National d'Etudes Spatiales (18 Av. BELIN - 31055 TOULOUSE Cédex)

search are developed in this book. Covering radius (G. Cohen/N.J.A. Sloane/A.C. Lobstein, H.F. Mattson Jr., L. Huguet/M. Griera), constructions/automorphisms of codes (J.A. Thiong-Ly, J.L. Dornstetter, D.A. Leonard/C.A. Rodger, B. Courteau/J. Goulet), practical aspects of codes (M.C. Gennero, G.L. Feng/K.K. Tzeng), polynomials (P.Piret, D. Lugiez, O. Moreno de Ayala), applied algebra (H.M. Möller/F. Mora, L. Beneteau/J. Lacaze, A. Astie-Vidal/J. Chifflet), cryptography (P. Camion), computer algebra (J. Calmet, J. Calmet/M. Bergman).

• AAECC Conferences essentially deal with Applied Algebra, Algorithmic and Error Correcting Codes.
The future scheduled AAECC conferences are :

- AAECC-3 (1985, Grenoble (F), Prof. J. Calmet)
- AAECC-4 (1986, Karlsruhe (D), Prof. Dr. T. Beth)
- AAECC-5 (1987, Barcelona (SP), Dr. L. Huguet)
- AAECC-6 (1988, Pisa (I), Prof. A. Miola)
- AAECC-7 (1989, Toulouse (F), Prof. A. Poli)
- AAECC-8 (1990, Yokohama (J), Prof. H. Imai)

We hope that AAECC Conferences, and particularly this Lecture Notes volume, will contribute to the important development of data transmissions.

Finally, a thank you to participants, authors, and also to Miss S. Watson (Springer Verlag Computer Science Editorial) for her patience and very kind help. A particular thanks to the series editors who have accepted this publication.

May 1986 Alain POLI

CONTENTS

ON ASSOCIATIVE ALGEBRAS OF MINIMAL RANK

Joos Heintz[1] and Jacques Morgenstern[2]

1) Consejo Nacional de Investigaciones Científicas y Técnicas (CONICET)
Universidad Nacional de La Plata
La Plata, Provincia Buenos Aires, Argentina
and
Johann Wolfgang Goethe - Universität, Fachbereich Mathematik
Robert Mayer - Strasse 6 - 10
D - 6000 Frankfurt/Main, FRG (mailing address)

2) Université de Nice, Institut des Mathématiques et Sciences Physiques
Parc Valrose
F - 06034 Nice Cedex, France
and
INRIA, Sophia Antipolis
F - 06560 Valbonne, France

1. Introduction

In the sequel let k be a field and A an associative k-algebra with unity, of finite dimension over k. We denote the radical of A, the maximal (two-sided) nilpotent ideal contained in A, by $\operatorname{rad} A$.

A quadratic algorithm (for A) is a finite family

$$\beta = ((u_\rho, v_\rho, w_\rho) \in (A \times A)^* \times (A \times A)^* \times A \; ; \; \rho = 1, \dots, R)$$

satisfying $\quad xy = \sum_{\rho=1}^{R} u_\rho(x,y)\, v_\rho(x,y)\, w_\rho \quad , \forall \; x,y \in A$.

(Here $(A \times A)^*$ denotes the dual space of the k-vector space $A \times A$.)

Special cases of quadratic algorithms are the bilinear algorithms (for A) which have the form $\beta = ((u_\rho, v_\rho, w_\rho) \in A^* \times A^* \times A \; ; \; \rho = 1, \dots, R)$ with

$$(1) \qquad xy = \sum_{\rho=1}^{R} u_\rho(x)\, v_\rho(y)\, w_\rho \quad , \forall \; x,y \in A .$$

(Note that $t := \Sigma\, u_\rho \otimes v_\rho \otimes w_\rho \in A^* \otimes_k A^* \otimes_k A$ is the tensor of the multiplication of the algebra A and hence doesn't depend on the particular algorithm β.)

For $\beta = ((u_\rho, v_\rho, w_\rho);\ \rho = 1,\dots,R)$ a quadratic or bilinear algorithm we call $L(\beta) := R$ the complexity of β. We define the following invariants of A :

$L(A) := \min\{L(\beta);\ \beta$ quadratic algorithm for $A\}$, the complexity of A and

$R(A) := \min\{L(\beta);\ \beta$ bilinear algorithm for $A\}$, the rank of A.

It is well known ([17]) that $L(A)$, the complexity of A, can be interpreted as the computational complexity of multiplying two generic elements of A.

Furthermore, we have $L(A) \leq R(A) \leq 2\,L(A)$.

So, for asymptotic complexity considerations, L and R are equivalent notions. This fact has widely been used for the construction of fast matrix multiplication algorithms (compare e.g. [16],[5],[15],[7]). Fast matrix multiplication and fast convolution algorithms are at the origin of the consideration of bilinear algorithms (compare e.g. [16], [17],[18],[19]).

The rank of an algebra appears to be closer related to the structure of A than its complexity. For this reason we focus our attention on the rank of algebras.

The starting point of our considerations is the following lower bound result for the complexity of associative algebras.

<u>Theorem 1</u> ([1]) $\quad L(A) \geq 2\dim_k A - \# M(A)$,

where $M(A) := \{m\ ;\ m$ maximal two-sided ideal of $A\}$ and $\# M(A)$ is its cardinality.

(In the case of $A := M_N(k)$, the algebra of $N \times N$ matrices over k, the result is due to [14].)

We will use the following notions :

<u>Definition 1</u> We say

(i) the complexity of A is minimal ($L(A)$ minimal)
iff $\quad L(A) = 2\dim_k A - \# M(A)$;

(ii) the rank of A is minimal ($R(A)$ minimal)
iff $\quad R(A) = 2\dim_k A - \# M(A)$.

Observations 1

1. We conjecture $L(A)$ minimal iff $R(A)$ minimal. This has been shown for A a division algebra ([8],[3]). In general, we only know that $R(A)$ minimal implies $L(A)$ minimal.
2. $M_2(k)$ is of minimal rank ([16]). We conjecture that $M_2(k)$ is the only matrix algebra of minimal rank.
3. Let k be infinite, X an indeterminate over k, and $F(X) \in k[X]$. Then $A := k[X]/(F)$ is of minimal rank ([19]).

We call the k-algebra A local if $A/\text{rad}\, A$ is a division algebra. We call A clean if for each maximal two-sided ideal m of A the k-algebra A/m is a division algebra. This is equivalent to saying that $A/\text{rad}\, A$ is a finite product of division algebras.

Note that A commutative implies A clean.

An important example of a non commutative clean k-algebra is $T_N(k)$, the algebra of the upper triangular $N \times N$ matrices over k.

We are considering the following class of k-algebras :

Definition 2 Let A be a clean k-algebra with $n := \dim_k A$ and such that $A/\text{rad}\, A$ is a p-fold product of division algebras. We say that A belongs to the class M_k (in symbols: $A \in M_k$), if there exists a pair of bases $\Sigma = ((x_1,\dots,x_n),(y_1,\dots,y_n))$ of A which satisfies the following properties :

(i) $x_\nu y_\mu \in k\, x_\nu + k\, y_\mu$ for each $1 \leq \nu,\mu \leq n$;

(ii) $x_\pi = y_\pi$ for $1 \leq \pi \leq p$, and $x_1,\dots,x_p$ are mutually orthogonal idempotents of A. Furthermore $1 = x_1 + \dots + x_p$, $x_\pi \cdot y_\mu \in k\, y_\mu$ and $x_\nu y_\pi \in k\, x_\nu$ for $1 \leq \pi \leq p$, $1 \leq \nu,\mu \leq n$.

A pair Σ of bases of A which satisfies (i) and (ii) is called a multiplicative pair (for short: an M-pair) of bases of A.

We remark that our notion of class M_k coincides with the one used in [6] in case of local k-algebras.

It is possible to characterize the class of commutative algebras of minimal rank over an infinite field k. In the case that k is algebraically closed, the result can be stated as follows :

Theorem 2 ([13]) Let k be algebraically closed and A be commutative. Then the following three conditions are equivalent :

(i) $R(A)$ minimal.

(ii) $A \in M_k$.

(iii) The radical of A has the form $\text{rad}\, A = (\omega_1,\ldots,\omega_m)$, where $\omega_i\,\omega_j = 0$ for $1 \leq i \neq j \leq m$.

Condition (iii) is a structural characterization of the class of commutative algebras of minimal rank over k . Our goal is to find such a structural characterization for any associative algebra of minimal rank. In this paper, we resolve this problem for the class of clean k-algebras in case k algebraically closed.

In the statement of our main result (Theorem 3) we use the following notions and notations :

Let A be a finite dimensional clean k-algebra over an algebraically closed field k .

For $\omega \in A$ we denote by $A\,\omega\,A$ the k-vector space of A generated by the products $a\,\omega\,b$, where $a,b \in A$. $A\,\omega\,A$ is the minimal two-sided ideal of A which contains ω .

Furthermore we write L and R for the following two-sided ideals of A : $L := \{x \in \text{rad}\, A\ ;\ x\,(\text{rad}\, A) = 0\}$ and $R := \{y \in \text{rad}\, A\ ;\ (\text{rad}\, A)\,y = 0\}$.

We have then

Theorem 3 In case k algebraically closed and A clean the following three conditions are equivalent :

(i) $R(A)$ minimal.

(ii) $A \in M_k$.

(iii) There exist $\omega_1,\ldots,\omega_m \in \text{rad}\, A$ such that

$$\text{rad}\, A = L + A\,\omega_1 A + \ldots + A\,\omega_m A = R + A\,\omega_1 A + \ldots + A\,\omega_m A \quad .$$

Here, in particular, the two-sided ideals $A\,\omega_i\,A$ are as k-vector subspaces of A generated by the products $\omega_i, \omega_i^2, \ldots, \omega_i^f, \ldots$, $f \in \mathbb{N}$, or equivalently: $A\,\omega_i\,A = \sum_{f \in \mathbb{N}} k\,\omega_i^f$.

Furthermore, $\omega_1,\ldots,\omega_m$ satisfy $\omega_i\,\omega_j = 0$ for $1 \leq i \neq j \leq m$.

Theorem 3 is known in special cases :
as Theorem 2 in case A commutative and Theorem 5 in [6] in case A local.

However, the origin of this kind of structure theorems is Theorem I.4 of [11], where a structural characterization of division algebras of minimal rank over an arbitrary but infinite field is given.

Observations 2

1. Principally, we are interested in the structural characterization (iii) of the complexity-theoretically defined class of clean k-algebras A of minimal rank. Motivated by the proof of Theorem 2, we insert the technical notion of class M_k (due to V. Strassen) between the structural notion (iii) and the complexity-theoretical notion (i) . This simplifies proofs sensibly. Under the assumptions of Theorem 3 (k algebraically closed and A clean) we will show in Section 2 : (i) $\Rightarrow$ (ii) (Proposition 1) , (ii) $\Rightarrow$ (iii) (Proposition 2) ; (iii) $\Rightarrow$ (i) (Proposition 3) follows in almost the same way as Proposition II.3 of [13].
2. Note that for A satisfying (iii) of Theorem 3 the following holds: Let $S \subset A$ any subalgebra with $A = S \oplus \text{rad}\, A$. Then $B := S[\omega_1, \ldots, \omega_m]$ is a commutative subalgebra of A of minimal rank with $A = L + B = R + B$. So, a clean k-algebra A of minimal rank is itself almost (i.e. modulo L or R) commutative.

The next corollary, due to a generalization [9] of Proposition 1 below, contains further consequences of Theorem 3 :

Corollary 1 Let k be an arbitrary but infinite field and A clean of minimal rank.

(i) rad A satisfies condition (iii) of Theorem 3 .

(ii) If a is a two-sided ideal of A then $R(A/a)$ is minimal .

As an easy application of Theorem 3 via Corollary 1 we obtain

Corollary 2 Let k be an arbitrary but infinite field. Denote by $T_N(k)$ the k-algebra of $N \times N$ upper triangular matrices. Then

$$R(T_N(k)) \text{ minimal} \quad \text{iff} \quad N = 2 .$$

Furthermore $R(T_3(k)) = 10$, the trivial 3×3 matrix multiplication algorithm being optimal in $T_3(k)$.

2. Proofs

In this section, we are going to sketch the proofs of three propositions which imply our main result Theorem 3. For general results about associative algebras we refer to [4].

In the proof of Proposition 2 we will make use of the following

Lemma 1 Let A a finite dimensional clean k-algebra with unity. Then each element $x \in A$ invertible in A is invertible in $k[x] \subset A$.

Proof Let $x \in A$ be invertible in A. Let $S := k[x]$. S is a finite dimensional, commutative k-algebra, hence artinian. So, $\text{rad}\, S$ is the set of all nilpotent elements of S. Since $\text{rad}\, A$ is a nilpotent ideal of A, this implies $S \cap \text{rad}\, A \subset \text{rad}\, S$.

We consider the embedding $S/(S \cap \text{rad}\, A) \hookrightarrow A/\text{rad}\, A$.

By $\text{rad}\, A = \bigcap_{\substack{m \text{ maximal} \\ \text{two-sided} \\ \text{ideal of } A}} m$ we have $A/\text{rad}\, A = \prod_{\substack{m \text{ maximal} \\ \text{two-sided} \\ \text{ideal of } A}} A/m$.

Since A is clean, $A/\text{rad}\, A$ is a finite product of division algebras. This implies that $S/(S \cap \text{rad}\, A)$ has no nilpotent element different from zero. Since $S/(S \cap \text{rad}\, A)$ is a commutative artinian k-algebra, we conclude $\text{rad}(S/(S \cap \text{rad}\, A)) = 0$, whence $\text{rad}\, S = S \cap \text{rad}\, A$.

Now, we consider $S/\text{rad}\, S$ as a subalgebra of $A/\text{rad}\, A$. Since x is invertible in A, the image of x in $A/\text{rad}\, A$ is in particular not a zero divisor. This implies that the image of x in $S/\text{rad}\, S$ doesn't divide zero neither.

Since S is commutative and artinian, it is a finite product of local k-algebras (compare e.g. [2], Theorem 8.7). $S/\text{rad}\, S$ is the product of the residue fields of these algebras.

Since the image of x in $S/\text{rad}\, S$ doesn't divide zero, its component in each residue field is different from zero. This implies that the component of x in each local algebra is invertible. Hence x is invertible in $S = k[x]$. □

As a corollary of Lemma 1 we obtain the following

Remark Let A a clean k-algebra. Let $x, y \in A$ with

$$xy = \xi x + \eta y \quad \text{, where} \quad \xi, \eta \in k \; , \; \xi \cdot \eta \neq 0 \; .$$

Then $k[x] = k[y]$.

Proof $xy = \xi x + \eta y$ implies $(x-\eta)(y-\xi) = \xi\eta$.
By hypothesis $\xi\eta \in k\setminus\{0\}$. So, $(x-\eta)$ and $(y-\xi)$ are invertible in A .
Since, by Lemma 1, $(y-\xi)^{-1} \in k[y-\xi] = k[y]$, we obtain
$x-\eta = \xi\eta(y-\xi)^{-1} \in k[y]$, whence $x \in k[y]$ and finally $k[x] \subset k[y]$.
Similarly, one proves $k[y] \subset k[x]$. □

Proposition 1 Let k be algebraically closed and A a clean k-algebra of minimal rank. Then $A \in M_k$.

(This proposition has been generalized to the case of an arbitrary but infinite field in [9].)

Proof Let $\bar{A} := A/\mathrm{rad}\, A$, $n := \dim_k A$, $p := \dim_k \bar{A}$.
Since A is clean, $\bar{A}$ is a finite product of division algebras. Furthermore, since k is algebraically closed and $p = \dim_k \bar{A}$, $\bar{A}$ is a p-fold product of the k-algebra k, so, without loss of generality, $\bar{A} = k^p$.
For $x \in A$ let $\bar{x} \in \bar{A}$ be its image under the canonical homomorphism $A \longrightarrow \bar{A}$. Since, by assumption, $\bar{A} = k^p$, $\bar{x}$ can be written as $\bar{x} = (x^{(1)},\ldots,x^{(p)})$ with $x^{(\pi)} \in k$, $\pi = 1,\ldots,p$.

Define the support of x as $\mathrm{supp}(x) := \{\pi ; 1 \leq \pi \leq p , x^{(\pi)} \neq 0\}$.

We note : $x \in \mathrm{rad}\, A$ iff $\mathrm{supp}(x) = \emptyset$
and x unity of A iff $\mathrm{supp}(x) = \{1,\ldots,p\}$.
Furthermore for $x,y \in A$ we have $\mathrm{supp}(x\cdot y) = \mathrm{supp}(x) \cap \mathrm{supp}(y)$.

We identify the space of k-linear endomorphisms of A with $A^* \otimes_k A$.
For $a \in A$ let $L_a , R_a \in A^* \otimes_k A$ the k-linear maps defined by
$L_a(x) := ax$, $R_a(x) := xa$ for any $x \in A$.

Let $t \in A^* \otimes_k A^* \otimes_k A$ be the tensor of the multiplication of A .

Since $n = \dim_k A$, $p = \dim_k \bar{A} = \# M(A)$, and $R(A)$ is minimal, we have $R(A) = 2n-p$.

So, let $\beta = ((u_\rho, v_\rho, w_\rho) \in A^* \times A^* \times A ; \rho = 1,\ldots,2n-p)$ with
$t = \sum_{\rho=1}^{2n-p} u_\rho \otimes v_\rho \otimes w_\rho$ an optimal bilinear algorithm for A .
The proof of Proposition 1 is based on methods inspired by [13], Theorem I.20. We make extensive use of the isotropy group of A ([10]), which we will apply subsequently to the algorithm β .

Since $id_A = R_1 = \sum_{\rho=1}^{2n-p} v_\rho(1)\; u_\rho \otimes w_\rho$ is regular, we may assume that

(2) $u_{n-p+1},\dots,u_{2n-p}$ form a base of A^* with dual base $x_{n-p+1},\dots,x_{2n-p} \in A$.

We may assume that

(3) $\bar{x}_{n-p+1},\dots,\bar{x}_n$ form a base of $\bar{A}$.

First, we show that we may assume without loss of generality

(4) $\mathrm{supp}(w_{n-p+1}) \neq \emptyset ,\dots, \mathrm{supp}(w_n) \neq \emptyset$.

Let the bilinear algorithm β be chosen such that $q := \#\{\pi ;\ 1 \leqq \pi \leqq p ,\ \mathrm{supp}(w_{n-p+\pi}) = \emptyset\}$ is minimal. We have to show $q = 0$.

Assume that this is not the case, and let $\mathrm{supp}(w_n) = \emptyset$ for instance. Since by (3) $\bar{x}_n \neq 0$, we have $x_n \notin \mathrm{rad}\, A$.

Then $x_n = R_1(x_n) = \sum_{\rho=1}^{n-p} u_\rho(x_n)\; v_\rho(1)\; w_\rho + v_n(1)\; w_n$ implies that there exists $1 \leqq \rho_o \leqq n-p$ with $u_{\rho_o}(x_n)\; v_{\rho_o}(1) \neq 0$, $w_{\rho_o} \notin \mathrm{rad}\, A$.

Without loss of generality, we may assume $\rho_o = 1$, so in particular

(5) $u_1(x_n) \neq 0$, $\mathrm{supp}(w_1) \neq \emptyset$.

For $\rho = 1,\dots,2n-p$ let
$$(\tilde{u}_\rho,\tilde{v}_\rho,\tilde{w}_\rho) := \begin{cases} (u_\rho,v_\rho,w_\rho) & \text{for } \rho \neq 1,n \\ (u_n,v_n,w_n) & \text{for } \rho = 1 \\ (u_1,v_1,w_1) & \text{for } \rho = n \end{cases} .$$

Then $\tilde{\beta} := ((\tilde{u}_\rho,\tilde{v}_\rho,\tilde{w}_\rho);\ \rho = 1,\dots,2n-p)$ is a new optimal algorithm for A . By (2) we have that

$$M := \{\tilde{u}_{n-p+1},\dots,\tilde{u}_{n-1},\tilde{u}_{n+1},\dots,\tilde{u}_{2n-p}\} = \{u_{n-p+1},\dots,u_{n-1},u_{n+1},\dots,u_{2n-p}\}$$

is a set of linearly independent elements of A^* , $x_n \in M^\perp$.
(Here $M^\perp$ denotes the orthogonal complement in A of the k-linear space generated by M in A^* .)

Furthermore, by (5), we have $\tilde{u}_n(x_n) = u_1(x_n) \neq 0$. Therefore $\tilde{u}_n$ is linearly independent of M and so $\tilde{u}_{n-p+1},\dots,\tilde{u}_{2n-p}$ form a base of A^*. Let $\tilde{x}_{n-p+1},\dots,\tilde{x}_{2n-p} \in A$ be its dual base.

We have

$$\tilde{x}_{n-p+\pi} = x_{n-p+\pi} - \frac{\tilde{u}_n(x_{n-p+\pi})}{\tilde{u}_n(x_n)}\, x_n \quad \text{for} \quad 1 \leq \pi \neq p \leq n \quad \text{and} \quad \tilde{x}_n = \frac{1}{\tilde{u}_n(x_n)}\, x_n .$$

Hence (3) implies that $\bar{\tilde{x}}_{n-p+1},\ldots,\bar{\tilde{x}}_n$ form a base of $\bar{A}$.

So, the bilinear algorithm β fulfills our assumptions (2),(3) we have made on β .

Since $\tilde{w}_{n-p+1} = w_{n-p+1},\ldots,\tilde{w}_{n-1} = w_{n-1}, \tilde{w}_n = w_1$ and $\operatorname{supp}(\tilde{w}_n) = \operatorname{supp}(w_1) \neq \emptyset$ by (5), we conclude $\#\{\ \pi\ ;\ 1 \leq \pi \leq p\ ,\ \operatorname{supp}(\tilde{w}_{n-p+\pi}) = \emptyset\ \} < q$, a contradiction.

Therefore, we shall assume from now on that our algorithm β satisfies (2),(3),(4).

Since, by (3), $\bar{x}_{n-p+1},\ldots,\bar{x}_n$ form a base of $\bar{A}$, there exist $\alpha_1,\ldots,\alpha_p \in k$ such that $\bar{1} = \sum_{\pi=1}^{p} \alpha_\pi\, \bar{x}_{n-p+\pi}$.

Let $x := \sum_{\pi=1}^{p} \alpha_\pi\, x_{n-p+\pi}$. x is a unity of A, so the mapping

$$L_x = \sum_{\rho=1}^{n-p} u_\rho(x)\, v_\rho \otimes w_\rho + \sum_{\pi=1}^{p} \alpha_\pi\, v_{n-p+\pi} \otimes w_{n-p+\pi} \quad \text{is regular.}$$

This implies

(6) $\quad v_1,\ldots,v_n$ form a base of A^* ,

(7) $\quad w_1,\ldots,w_n$ form a base of A ,

and $\alpha_1 \neq 0,\ldots,\alpha_p \neq 0$.

Scaling $u_{n-p+1},\ldots,u_n$ against $v_{n-p+1},\ldots,v_n$, we may assume $\alpha_1 = \ldots = \alpha_p = 1$ and hence $x = \sum_{\pi=1}^{p} x_{n-p+\pi}$.

For $\rho = 1,\ldots,2n-p$ let $\tilde{u}_\rho := u_\rho \circ R_x$, $\tilde{v}_\rho := v_\rho \circ L_{x^{-1}}$, $\tilde{w}_\rho := w_\rho$.

Then, for any $a,b \in A$ we have

$$\sum_{\rho=1}^{2n-p} \tilde{u}_\rho(a)\, \tilde{v}_\rho(b)\, w_\rho = \sum_{\rho=1}^{2n-p} u_\rho(a x)\, v_\rho(x^{-1} b)\, w_\rho = (a x)(x^{-1} b) = ab .$$

Hence $\tilde{\beta} := ((\tilde{u}_\rho,\tilde{v}_\rho,\tilde{w}_\rho)\,;\, \rho = 1,\ldots,2n-p)$ is an optimal algorithm for A .

Note that $\tilde{u}_{n-p+1} = u_{n-p+1} \circ R_x ,\ldots, \tilde{u}_{2n-p} = u_{2n-p} \circ R_x$ form a base of A^* with dual base $\tilde{x}_{n-p+1} = x_{n-p+1}\, x^{-1} ,\ldots, \tilde{x}_{2n-p} = x_{2n-p}\, x^{-1}$.

This implies that $\bar{\tilde{x}}_{n-p+1},\ldots,\bar{\tilde{x}}_n$ form a base of $\bar{A}$. Furthermore $\tilde{w}_{n-p+1} = w_{n-p+1} ,\ldots, \tilde{w}_n = w_n$ implies $\operatorname{supp}(\tilde{w}_{n-p+1}) \neq \emptyset ,\ldots, \operatorname{supp}(\tilde{w}_n) \neq \emptyset$.

Therefore $\tilde{\beta}$ fulfills the assumptions (2),(3),(4) we made on β.

By $x = \sum_{\pi=1}^{p} x_{n-p+\pi}$ we have $1 = \sum_{\pi=1}^{p} x_{n-p+\pi} x^{-1} = \sum_{\pi=1}^{p} \tilde{x}_{n-p+\pi}$.
Hence substituting β by $\tilde{\beta}$ we may assume that β satisfies

(8) $$1 = \sum_{\pi=1}^{p} x_{n-p+\pi} .$$

Let $U := \{u_{n+1},\dots,u_{2n-p}\}^{\perp}$, $V := \{v_1,\dots,v_{n-p}\}^{\perp}$.

Since, by (2) and (6), $u_{n-p+1},\dots,u_{2n-p}$ and $v_1,\dots,v_n$ form bases of A^*, we have $\dim U = \dim V = p$.

Furthermore, we have by (8) $1 \in U = \sum_{\pi=1}^{p} k\, x_{n-p+\pi}$.

For $y \in V$ we get from (1)

(9) $$x_{n-p+\pi}\, y = v_{n-p+\pi}(y)\, w_{n-p+\pi} \quad \text{for} \quad 1 \leq \pi \leq p .$$

So, by (8), $y = 1 \cdot y = \sum_{\pi=1}^{p} x_{n-p+\pi} \cdot y = \sum_{\pi=1}^{p} v_{n-p+\pi}(y)\, w_{n-p+\pi} \subset \sum_{\pi=1}^{p} k\, w_{n-p+\pi}$.

From $\dim V = p$ we conclude $V = \sum_{\pi=1}^{p} k\, w_{n-p+\pi}$.

Now, $w_{n-p+1},\dots,w_n \in V$, (8),(9) imply for $1 \leqq \pi \leqq p$

$$w_{n-p+\pi} = 1 \cdot w_{n-p+\pi} = \sum_{\pi'=1}^{p} x_{n-p+\pi'} \cdot w_{n-p+\pi} = \sum_{\pi'=1}^{p} v_{n-p+\pi'}(w_{n-p+\pi})\, w_{n-p+\pi'} .$$

Since, by (7), $w_1,\dots,w_n$ form a base of A, this implies $v_{n-p+\pi'}(w_{n-p+\pi}) = \delta_{\pi'\pi}$ for $1 \leqq \pi',\pi \leqq p$ ($\delta_{\pi'\pi}$ being the Kronecker symbol).

By (9) we have in particular

(10) $$x_{n-p+\pi'}\, w_{n-p+\pi} = \delta_{\pi'\pi}\, w_{n-p+\pi} ,$$

whence

(11) $$\operatorname{supp}(w_{n-p+\pi}) \subset \operatorname{supp}(x_{n-p+\pi}) .$$

For $1 \leqq \pi \leqq p$ we consider $w_{n-p+\pi}$. Since, by (4), $\operatorname{supp}(w_{n-p+\pi}) \neq \emptyset$, there exists $1 \leqq \pi' \leqq p$ with $w_{n-p+\pi}^{(\pi')} \neq 0$.

We are going to show $\operatorname{supp}(w_{n-p+\pi}) = \{\pi'\}$.

Since $\bar{x}_{n-p+1},\dots,\bar{x}_n$ is a base of $\bar{A}$, we can choose $\alpha_1,\dots,\alpha_p \in k$ such that $\operatorname{supp}(z) = \{\pi'\}$ for $z := \sum_{\theta=1}^{p} \alpha_\theta\, x_{n-p+\theta}$.

We have by (10) $z\, w_{n-p+\pi} = \sum_{\theta=1}^{p} \alpha_\theta\, x_{n-p+\theta}\, w_{n-p+\pi} = \alpha_\pi\, w_{n-p+\pi}$.

Now $\{\pi'\} = \mathrm{supp}(z) \cap \mathrm{supp}(w_{n-p+\pi}) = \mathrm{supp}(z\, w_{n-p+\pi})$ implies $\alpha_\pi \neq 0$ and finally $\mathrm{supp}(w_{n-p+\pi}) = \{\pi'\}$.

Let ψ be the map $\psi : \{1,\dots,p\} \longrightarrow \{1,\dots,p\}$ given by $\mathrm{supp}(w_{n-p+\pi}) = \{\psi(\pi)\}$, $1 \leq \pi \leq p$. We claim that ψ is injective and hence bijective.

Otherwise, let $1 \leqq \pi \neq \pi' \leqq p$ such that $\psi(\pi) = \psi(\pi')$.
This means $\mathrm{supp}(w_{n-p+\pi}) = \mathrm{supp}(w_{n-p+\pi'}) = \{\theta\}$, where $\theta = \psi(\pi) = \psi(\pi')$.
Since, by (11), $\mathrm{supp}(w_{n-p+\pi}) \subset \mathrm{supp}(x_{n-p+\pi})$ this implies $\theta \in \mathrm{supp}(x_{n-p+\pi})$, i.e. $x^{(\theta)}_{n-p+\pi} \neq 0$.

Furthermore, we have $w^{(\theta)}_{n-p+\pi'} \neq 0$, whence $x^{(\theta)}_{n-p+\pi}\, w^{(\theta)}_{n-p+\pi'} \neq 0$, which contradicts (10), namely $x_{n-p+\pi}\, w_{n-p+\pi'} = \delta_{\pi\pi'}\, w_{n-p+\pi'} = 0$.

The bijectivity of ψ implies that $y := \sum_{\pi=1}^{p} w_{n-p+\pi} \in V$ is a unity.
Substituting $\tilde{\beta} = ((\tilde{u}_\rho, \tilde{v}_\rho, \tilde{w}_\rho) = (u_\rho, v_\rho \circ R_y, w_\rho\, y^{-1})\, ;\, \rho = 1,\dots,2n-p)$ for β, one sees, as before, that one may assume

$$(12) \qquad 1 = y = \sum_{\pi=1}^{p} w_{n-p+\pi} .$$

By (6) $v_1,\dots,v_n$ form a base of A^*.
Let $y_1,\dots,y_n \in A$ its dual base. For $\pi = 1,\dots,p$ we have by (1) and (8) $y_{n-p+\pi} = 1 \cdot y_{n-p+\pi} = \sum_{\pi'=1} x_{n-p+\pi'}\, y_{n-p+\pi} = w_{n-p+\pi}$.
Therefore, (12) implies

$$(13) \qquad 1 = \sum_{\pi=1}^{p} w_{n-p+\pi} = \sum_{\pi=1}^{p} y_{n-p+\pi} .$$

Similarly, using (1) and (13), one shows

$$x_{n-p+\pi} = x_{n-p+\pi} \cdot 1 = \sum_{\pi'=1}^{p} x_{n-p+\pi}\, y_{n-p+\pi'} = w_{n-p+\pi} .$$

Now, (1) and (13) imply for $\nu = p,\dots,n$

$$x_{n-p+\nu} = x_{n-p+\nu} \cdot 1 = \sum_{\pi=1}^{p} x_{n-p+\nu}\, y_{n-p+\pi} = \sum_{\pi=1}^{p} v_{n-p+\nu}(y_{n-p+\pi})\, w_{n-p+\nu}$$

$$= v_{n-p+\nu}(1)\, w_{n-p+\nu} ,$$

where $v_{n-p+\nu}(1) \neq 0$.

Scaling $v_{n-p+\nu}$ against $w_{n-p+\nu}$ for $\nu > p$ and taking into account $x_{n-p+1} = w_{n-p+1}, \ldots, x_n = w_n$, we may assume $x_{n-p+\nu} = w_{n-p+\nu}$, whence

(14) $\quad x_{n-p+1} = w_{n-p+1}, \ldots, x_{2n-p} = w_{2n-p}$.

Similarly, for $\mu = 1,\ldots,n-p$

$$y_\mu = 1\cdot y_\mu = \sum_{\pi=1}^{p} x_{n-p+\pi}\, y_\mu = \sum_{\pi=1}^{p} u_\mu(x_{n-p+\pi})\, y_\mu = u_\mu(1)\, w_\mu ,$$

where $u_\mu(1) \neq 0$.

Scaling u_μ against w_μ for $\mu \leq n-p$ and taking into account $y_{n-p+1} = w_{n-p+1}, \ldots, y_n = w_n$, we may assume $y_\mu = w_\mu$, whence

(15) $\quad y_1 = w_1, \ldots, y_n = w_n$.

Now, (1),(14),(15) imply for $1 \leq \nu,\mu \leq n$

(16) $$w_{n-p+\nu}\, w_\mu = x_{n-p+\nu}\, y_\mu = u_\mu(x_{n-p+\nu})\, w_\mu + v_{n-p+\nu}(y_\mu)\, w_{n-p+\nu} \in k\, w_\mu + k\, w_{n-p+\nu} .$$

Furthermore, we have by (1),(14),(15) for $1 \leq \pi,\pi' \leq p$

(17) $$w_{n-p+\pi}\, w_{n-p+\pi'} = x_{n-p+\pi}\, y_{n-p+\pi'} = \delta_{\pi\pi'}\, w_{n-p+\pi} .$$

Hence $w_{n-p+1},\ldots,w_n$ are mutually orthogonal idempotents of A.

From (1),(14),(15) we conclude for any $1 \leq \pi \leq p$, $1 \leq \nu,\mu \leq n$

(18) $$w_{n-p+\pi}\, w_\mu = x_{n-p+\pi}\, y_\mu = u_\mu(x_{n-p+\pi})\, w_\mu \in k\, w_\mu \quad \text{and}$$
$$w_{n-p+\nu}\, w_{n-p+\pi} = x_{n-p+\nu}\, y_{n-p+\pi} = v_{n-p+\nu}(y_{n-p+\pi})\, w_{n-p+\nu} \in k\, w_{n-p+\nu} .$$

Now, (13),(14),(15),(16),(17),(18) imply that

$$\Sigma = ((w_{n-p+1},\ldots,w_{2n-p}),(w_{n-p+1},\ldots,w_n, w_1,\ldots,w_{n-p}))$$

is an M-pair of bases of A.

Therefore $A \in M_k$. □

<u>Proposition 2</u> Let k be algebraically closed, and let A be a clean k-algebra with $A \in M_k$.
Then A has the following property:

(*)
There exist $\omega_1, \ldots, \omega_m \in \operatorname{rad} A$ such that
$\operatorname{rad} A = L + A\omega_1 A + \ldots + A\omega_m A = R + A\omega_1 A + \ldots + A\omega_m A$ holds.
Furthermore, we have $A\omega_i A = \sum_{f \in \mathbb{N}} k\, \omega_i^f$, $\omega_i^2 \neq 0$, and
$\omega_i \omega_j = 0$ for $1 \leq i \neq j \leq m$.

(Here L and R denote the two-sided ideals of A
$L := \{ x \in \operatorname{rad} A \,;\, x(\operatorname{rad} A) = 0 \}$ and $R := \{ y \in \operatorname{rad} A \,;\, (\operatorname{rad} A)y = 0 \}$.)

<u>Proof</u> Let k be algebraically closed and A be a clean k-algebra with $A \in M_k$.

We maintain the notations of the proof of Proposition 1, in particular: $\bar{A} := A/\operatorname{rad} A$, $n := \dim_k A$, $p := \dim_k \bar{A}$.
Identifying $\bar{A} = k^p$, we write $\bar{x} = (x^{(1)}, \ldots, x^{(p)})$ with $x^{(\pi)} \in k$, $1 \leq \pi \leq p$, for the image of x in $\bar{A}$.
Furthermore, we write $\operatorname{supp}(x) := \{ \pi \,;\, 1 \leq \pi \leq p \,,\, x^{(\pi)} \neq 0 \}$.
During the proof, let X, Y be indeterminates over k .

Since, by hypothesis, $A \in M_k$, there exists an M-pair of bases of A
$\Sigma = ((x_1, \ldots, x_n), (y_1, \ldots, y_n))$.

In a first step of the proof, we shall simplify Σ .
From the Definition 2(ii) it is immediate, that $\bar{x}_1 = \bar{y}_1, \ldots, \bar{x}_p = \bar{y}_p$ is a system of mutually orthogonal idempotents of $\bar{A} = k^p$, none of them being zero, hence, in particular, they form a base of $\bar{A}$.
So, we may assume without loss of generality

(19) $\operatorname{supp}(x_1) = \operatorname{supp}(y_1) = \{1\} , \ldots , \operatorname{supp}(x_p) = \operatorname{supp}(y_p) = \{p\}$.

Furthermore, we conclude from $x_\pi y_\mu \in k\, y_\mu$, $x_\nu y_\pi \in k\, x_\nu$, and $x_\pi = y_\pi$ idempotent (Definition 2 (ii)) $x_\pi y_\mu \in \{0, y_\mu\}$, $x_\nu y_\pi \in \{0, x_\nu\}$ for any $1 \leq \pi \leq p$, $1 \leq \nu, \mu \leq n$.

For $p < \sigma, \tau \leq n$ let $x^*_\sigma := x_\sigma - \sum_{1 \leq \pi \leq p} x^{(\pi)}_\sigma x_\pi$, $y^*_\tau := y_\tau - \sum_{1 \leq \pi \leq p} y^{(\pi)}_\tau y_\pi$.
Then, we have by (19)

(20) $x^*_{p+1}, \ldots, x^*_n \,,\, y^*_{p+1}, \ldots, y^*_n \in \operatorname{rad} A$.

Using Definition 2 (i) and (ii), we obtain

$$x^*_\sigma y^*_\tau = x_\sigma y_\tau - \sum_{1\le\pi\le p} x_\sigma^{(\pi)} x_\pi y_\tau - \sum_{1\le\pi\le p} y_\tau^{(\pi)} x_\sigma y_\pi + \sum_{1\le\pi,\pi'\le p} x_\sigma^{(\pi)} y_\tau^{(\pi')} x_\pi y_{\pi'}$$

$$= \xi x_\sigma + \eta y_\tau + \sum_{1\le\pi\le p} \alpha_\pi x_\pi = \xi x^*_\sigma + \eta y^*_\tau + \sum_{1\le\pi\le p} \beta_\pi x_\pi ,$$

where $\xi,\eta,\alpha_1,\dots,\alpha_p, \beta_1,\dots,\beta_p \in k$ are suitably chosen.

Since, by (20), $x^*_\sigma, y^*_\tau \in \operatorname{rad} A$ this implies $\sum_{1\le\pi\le p} \beta_\pi x_\pi \in \operatorname{rad} A$, whence $\sum_{1\le\pi\le p} \beta_\pi \bar{x}_\pi = 0$.

But, $\bar{x}_1,\dots,\bar{x}_p$ form a base of $\bar{A}$, so $\beta_1 = \dots = \beta_p = 0$.

Therefore, we obtain finally $x^*_\sigma y^*_\tau = \xi x^*_\sigma + \eta y^*_\tau$, whence

(21) $\quad x^*_\sigma y^*_\tau \in k\, x^*_\sigma + k\, y^*_\tau \quad$ for any $p < \sigma,\tau \le n$.

Next, we are going to show

(22) $\quad x_\pi y^*_\tau \in k\, y^*_\tau$ and $x^*_\sigma y_\pi \in k\, x^*_\sigma$ for any $1 \le \pi \le p$, $p < \sigma,\tau \le n$.

Let $1 \le \pi \le p$ and $p < \tau \le n$. From Definition 2 (ii) we have

$x_\pi y^*_\tau = x_\pi y_\tau - \sum_{1\le\pi'\le p} y_\tau^{(\pi')} x_\pi y_{\pi'} = \eta\, y_\tau - y_\tau^{(\pi)} y_\pi$, where $\eta \in \{0,1\}$.

Note $x_\pi y^*_\tau \in \operatorname{rad} A$ by (20).

So, if $\eta = 0$, we conclude $y_\tau^{(\pi)} = 0$. Hence $x_\pi y^*_\tau = 0$ in this case.

Let $\eta = 1$. For any $1 \le \pi' \ne \pi \le p$ we have

$$0 = (x_\pi y^*_\tau)^{(\pi')} = y_\tau^{(\pi')} - y_\tau^{(\pi)} y_\pi^{(\pi')} = y_\tau^{(\pi')} .$$

This implies $y^*_\tau = y_\tau - \sum_{1\le\pi'\le p} y_\tau^{(\pi')} y_{\pi'} = y_\tau - y_\tau^{(\pi)} y_\pi$ and finally

$$x_\pi y^*_\tau = \eta\, y_\tau - y_\tau^{(\pi)} y_\pi = y_\tau - y_\tau^{(\pi)} y_\pi = y^*_\tau .$$

In both cases we have $x_\pi y^*_\tau \in k\, y^*_\tau$.

The second assertion of (22) follows in the same way.

Now, (21) and (22) imply that

$$\Sigma^* = ((x_1,\dots,x_p,x^*_{p+1},\dots,x^*_n),(y_1,\dots,y_p,y^*_{p+1},\dots,y^*_n))$$

is a M-pair of bases of A.

Therefore, taking into account (20), we may assume without loss of generality that Σ satisfies

(23) $\quad x_{p+1},\dots,x_n ,\ y_{p+1},\dots,y_n \in \operatorname{rad} A$.

Moreover, we have

(24) $\quad \operatorname{rad} A = \sum_{p<\sigma\leq n} k\, x_\sigma = \sum_{p<\tau\leq n} k\, y_\tau$.

To show (24) let $x \in \operatorname{rad} A$. Then $x = \sum_{1\leq\pi\leq p} \xi_\pi x_\pi + \sum_{p<\sigma\leq n} \xi_\sigma x_\sigma$.

From (23) we conclude $0 = \bar{x} = \sum_{1\leq\pi\leq p} \xi_\pi \bar{x}_\pi$, and, since $\bar{x}_1,\ldots,\bar{x}_p$ form a base of $\bar{A}$, $\xi_1 = \ldots = \xi_p = 0$. Therefore $x = \sum_{p<\sigma\leq n} \xi_\sigma x_\sigma \in \sum_{p<\sigma\leq n} k\, x_\sigma$.

Now $\operatorname{rad} A = \sum_{p<\sigma\leq n} k\, x_\sigma$ follows by (23).

The second assertion of (24) is shown in the same way.

Next, we are going to describe $\operatorname{rad} A$ in terms of the M-pair Σ of bases of A , Σ satisfying (24).

Let $\Gamma := \{\sigma\,;\, p<\sigma\leq n\,,\, x_\sigma^2 \neq 0\,,\, \exists\, \tau\,,\, p<\tau\leq n\,,\, \exists\, \xi,\eta \in k\setminus\{0\} : x_\sigma\, y_\tau = \xi\, x_\sigma + \eta\, y_\tau\}$,

$\Delta := \{\tau\,;\, p<\tau\leq n\,,\, y_\tau^2 \neq 0\,,\, \exists\, \sigma\,,\, p<\sigma\leq n\,,\, \exists\, \xi,\eta \in k\setminus\{0\} : x_\sigma\, y_\tau = \xi\, x_\sigma + \eta\, y_\tau\}$.

First, note the following:

(25)
Let $\gamma \in \Gamma$ and $p < \tau \leq n$.
Then $x_\gamma y_\tau = \xi x_\gamma + \eta y_\tau$ with $\xi,\eta \in k\setminus\{0\}$ implies $\tau \in \Delta$.
Let $p < \sigma \leq n$ and $\delta \in \Delta$.
Then $x_\sigma y_\delta = \xi x_\sigma + \eta y_\delta$ with $\xi,\eta \in k\setminus\{0\}$ implies $\sigma \in \Gamma$.

We show the first assertion of (25).

Let $\gamma \in \Gamma$ and $p < \tau \leq n$ with $x_\gamma y_\tau = \xi\, x_\gamma + \eta\, y_\tau$, where $\xi,\eta \in k\setminus\{0\}$.

Assume $\tau \notin \Delta$.

Then $y_\tau^2 = 0$. This implies $0 = x_\gamma y_\tau^2 = \xi\, x_\gamma y_\tau + \eta\, y_\tau^2 = \xi\, x_\gamma y_\tau$. Therefore, by $\xi \neq 0$, $x_\gamma\, y_\tau = 0$. Now $\xi\, x_\gamma = -\eta\, y_\tau$, whence $\xi\, x_\gamma^2 = -\eta\, x_\gamma\, y_\tau = 0$, which contradicts $x_\gamma^2 \neq 0$.

We will also make use of the following fact:

(26) $\quad k[x_\sigma] = k[y_\tau] \quad$ or $\quad x_\sigma\, y_\tau = 0 \qquad$ for $p < \sigma,\tau \leq n$.

We show (26). Let $p < \sigma,\tau \leq n$.

Then $x_\sigma y_\tau = \xi\, x_\sigma + \eta\, y_\tau$ for $\xi,\eta \in k\setminus\{0\}$ or $x_\sigma y_\tau \in k\, x_\sigma$ or $x_\sigma y_\tau \in k\, y_\tau$.

In the first case we have $k[x_\sigma] = k[y_\tau]$ by Remark .

Now, we consider the second case, namely $x_\sigma y_\tau \in k\, x_\sigma$.

Let $x_\sigma y_\tau = \xi x_\sigma$, where $\xi \in k$.
Thus, for any $f \in \mathbb{N}$ we have $x_\sigma y_\tau^f = \xi^f x_\sigma$. Since, by (23), $y_\tau \in \operatorname{rad} A$, y_τ is nilpotent. Choose $f \in \mathbb{N}$ such that $y_\tau^f = 0$. Then, we have $0 = x_\sigma y_\tau^f = \xi^f x_\sigma$. Since $x_\sigma \neq 0$, this implies $\xi = 0$. Therefore $x_\sigma y_\tau = 0$.
In the same way, one shows $x_\sigma y_\tau = 0$ in the third case.

From (24) we conclude : for $p < \sigma, \tau \leq n$

$$(27) \quad \begin{array}{lll} \sigma \notin \Gamma & \text{implies} & x_\sigma \in L \ , \\ \tau \notin \Delta & \text{implies} & y_\tau \in R \ . \end{array}$$

We prove the first assertion of (27).
Let $p < \sigma \leq n$, $\sigma \notin \Gamma$. We show that for any $p < \tau \leq n$ the assertion $x_\sigma y_\tau = 0$ holds. Thus, by (24), we conclude $x_\sigma \in L$.
Let $p < \tau \leq n$ be arbitrary. First, we consider the case that there exist $\xi, \eta \in k \setminus \{0\}$ with $x_\sigma y_\tau = \xi x_\sigma + \eta y_\tau$. By assumption, $\sigma \notin \Gamma$. So, we have $x_\sigma^2 = 0$. Therefore $0 = x_\sigma^2 y_\tau = \xi x_\sigma^2 + \eta x_\sigma y_\tau = \eta x_\sigma y_\tau$.
From $\eta \neq 0$ we conclude $x_\sigma y_\tau = 0$.
Now, consider the cases, where $x_\sigma y_\tau \in k x_\sigma$ or $x_\sigma y_\tau \in k y_\tau$.
In both cases we conclude, as in the proof of (26), $x_\sigma y_\tau = 0$ by the nilpotency of $\operatorname{rad} A$.

Now, (24) and (27) imply that $\operatorname{rad} A$ has the following form :

$$(28) \quad \operatorname{rad} A = L + \sum_{\sigma \in \Gamma} k x_\sigma = R + \sum_{\tau \in \Delta} k y_\tau \ .$$

For $p < \sigma, \tau \leq n$ we write

$$A_\sigma := \sum_{f \in \mathbb{N}} k x_\sigma^f = \{ F(x_\sigma) \ ; \ F(X) \in k[X] \ , \ F(0) = 0 \} \ ,$$
$$B_\tau := \sum_{f \in \mathbb{N}} k y_\tau^f = \{ G(y_\tau) \ ; \ G(Y) \in k[Y] \ , \ G(0) = 0 \} \ .$$

With this notation we conclude by (23)

$$(29) \quad A_\sigma , B_\tau \subset \operatorname{rad} A \quad \text{and} \quad A_\sigma = B_\tau \quad \text{if} \quad k[x_\sigma] = k[y_\tau] \ .$$

Therefore, taking into account Remark, (25), (28), and (29),

$$\operatorname{rad} A = L + a = R + a \ ,$$

where a is the k-vector space $a := \sum_{\sigma \in \Gamma} A_\sigma = \sum_{\tau \in \Delta} B_\tau$.

We are going to show that a is a two-sided ideal of A of a very particular form.

Let $\sigma \in \Gamma$, $x \in A_\sigma$. Definition 2 (ii) and (26) imply $x A \subset A_\sigma$. Therefore $A_\sigma A = A_\sigma$. Analogously for $\tau \in \Delta$, $A B_\tau = B_\tau$. So, by Remark, (25), and (29), we obtain for $p < \sigma, \tau \leq n$

(30) A_σ and B_τ are the two-sided ideals

$$A_\sigma = \sum_{f \in \mathbb{N}} k x_\sigma^f = A x_\sigma A \quad , \quad B_\tau = \sum_{f \in \mathbb{N}} k y_\tau^f = A y_\tau A \quad .$$

(30) implies now that $\mathfrak{a}$ is a two-sided ideal of A.

Let $\Gamma \dot{\cup} \Delta$ be the disjoint union of Γ and Δ.

Following an idea of V. Strassen (see [12]), we define a bipartite graph (Z,E), where the set of vertices is $Z := \Gamma \dot{\cup} \Delta$ and the set of edges $E \subset \Gamma \times \Delta$ is given by $(\sigma,\tau) \in E$ iff $k[x_\sigma] = k[y_\tau]$.

σ, σ' are in the same connected component of (Z,E) iff there exists a finite sequence $\sigma_1, \tau_1, \sigma_2, \tau_2, \ldots, \tau_{s-1}, \sigma_s$ with $\sigma = \sigma_1, \sigma_2, \ldots, \sigma' = \sigma_s \in \Gamma$, $\tau_1, \ldots, \tau_{s-1} \in \Delta$ and

$$k[x_\sigma] = k[x_{\sigma_1}] = k[y_{\tau_1}] = k[x_{\sigma_2}] = k[y_{\tau_2}] = \ldots = k[y_{\tau_{s-1}}] = k[x_{\sigma_s}] = k[x_{\sigma'}] \; .$$

In particular, we have by (29):

(31) $A_\sigma = A_{\sigma'}$ if σ, σ' are in the same connected component of (Z,E).

Now, let $\sigma, \sigma' \in \Gamma$ be in different connected components of (Z,E). By (25) and Remark, there exist $\tau, \tau' \in \Delta$ with $k[x_\sigma] = k[y_\tau]$ and $k[x_{\sigma'}] = k[y_{\tau'}]$. Since σ and σ' are not connected, we have $k[x_\sigma] \neq k[y_{\tau'}]$. Therefore, by (26), $x_\sigma y_{\tau'} = 0$.

$k[x_{\sigma'}] = k[y_{\tau'}]$ implies that there exists a polynomial $G(Y) \in k[Y]$, such that $x_{\sigma'} = G(y_{\tau'})$.

Since $x_{\sigma'}, y_{\tau'} \in \operatorname{rad} A$ by (23), we have $G(0) = 0$, i.e. $G(Y) = Y\,G^*(Y)$, where $G^*(Y) \in k[Y]$.

Now, $x_\sigma y_{\tau'} = 0$ implies $x_\sigma x_{\sigma'} = x_\sigma G(y_{\tau'}) = x_\sigma y_{\tau'} G^*(y_{\tau'}) = 0$.

We have shown

(32) $x_\sigma x_{\sigma'} = 0$ if $\sigma, \sigma' \in \Gamma$ are in different connected components of (Z,E).

Without restriction of generality, we may assume that $p+1,\dots,p+m$, with $m \leq n-p$, is a complete system of representants for the connected components of (Z,E) through Γ.

Let $\omega_1 := x_{p+1}, \dots, \omega_m := x_{p+m}$.

Then, by (30) and (31) $\mathfrak{a} = A_{p+1} + \dots + A_{p+m} = A\,\omega_1 A + \dots + A\,\omega_m A$,

where $A\,\omega_i\,A = \sum_{f \in \mathbb{N}} k\,\omega_i^f$ for $1 \leq i \leq m$.

Furthermore, we have for $1 \leq i \neq j \leq m$ $p+i \in \Gamma$ and therefore

$\omega_i^2 = x_{i+p}^2 \neq 0$, and by (32) $\omega_i\,\omega_j = x_{p+i}\,x_{p+j} = 0$.

We can resume our results as follows:

> There exist elements $\omega_1,\dots,\omega_m \in \operatorname{rad} A$ such that
>
> $A\,\omega_i\,A = \sum_{f \in \mathbb{N}} k\,\omega_i^f$, $\omega_i^2 \neq 0$, $\omega_i\,\omega_j = 0$ for $1 \leq i \neq j \leq m$
>
> and such that
>
> $\operatorname{rad} A = L + A\,\omega_1 A + \dots + A\,\omega_m A = R + A\,\omega_1 A + \dots + A\,\omega_m A$.

This finishes the proof of Proposition 2. □

Lemma 2 (see Observation 2.2) Let k be algebraically closed and let A be clean, satisfying Theorem 3 (iii). Thus A satisfies (*).
Let S be any (semisimple) subalgebra of A such that $A = S \oplus \operatorname{rad} A$.
Then $B := S[\omega_1,\dots,\omega_m]$ is a commutative subalgebra of A with $B \in M_k$ and $A = L + B = R + B$.

Proof Let $\bar{A} := A/\operatorname{rad} A$, $p := \dim_k \bar{A}$, and let S be a subalgebra of A with $A = S \oplus \operatorname{rad} A$.

Since $S \cong \bar{A} \cong k^p$, there exist mutually orthogonal idempotents $e_1,\dots,e_p \in S$ forming a base of S.

We are going to show that

(32) for any $\omega \in \{\omega_1,\dots,\omega_m\}$ and any idempotent e of A
$e\omega = \omega e \in \{0,\omega\}$

holds.

Then $e_1,\dots,e_p$ and $\omega_1,\dots,\omega_m$ commute.
Since, by (*), the elements $\omega_1,\dots,\omega_m$ annihilate each other, they commute.

Therefore, $\mathcal{B} := S[\omega_1,\ldots,\omega_m]$ is a commutative k-algebra.

We have $\mathcal{B} \in M_k$ by Theorem 2.

Furthermore, (*) implies $A = L + \mathcal{B} = R + \mathcal{B}$.

Now, let $\omega \in \{\omega_1,\ldots,\omega_m\}$ and e any idempotent of A.

By (*) we have $e\omega = \xi\omega + \omega^f F(\omega)$, where $\xi \in k$, $f \in \mathbb{N}$, $f \geqq 2$, $F(X) \in k[X]$. We can choose f such that $\omega^f \neq 0$ and $F(0) \neq 0$ or such that $\omega^f F(\omega) = 0$ holds.

In case $\omega^f \neq 0$ and $F(0) \neq 0$ we have

$$\xi\omega + \omega^f F(\omega) = e\omega = e^2\omega = e(\xi\omega + \omega^f F(\omega)) =$$
$$= \xi e\omega + e\omega^f F(\omega) = \xi^2\omega + 2\xi\omega^f F(\omega) + \omega^{2f-1} F(\omega)^2 .$$

Comparing powers of ω, we conclude from $\omega^f \neq 0$, $F(0) \neq 0$, $f \geqq 2$, that $2\xi = 0$.

Therefore, $\xi\omega + \omega^f F(\omega) = \xi^2\omega + \omega^{2f-1} F(\omega)^2$, which contradicts $\omega^f \neq 0$, $F(0) \neq 0$, and $f \geqq 2$.

Therefore, $\omega^f F(\omega) = 0$, whence $e\omega = \xi\omega$.

Since e is idempotent, we have $\xi \in \{0,1\}$, i.e. $e\omega \in \{0,\omega\}$.
Analogously, $\omega e \in \{0,\omega\}$.

If $e\omega \neq \omega e$, we may assume $e\omega = \omega$, $\omega e = 0$.

Then $\omega^2 = \omega(e\omega) = (\omega e)\omega = 0$. But, since $\omega \in \{\omega_1,\ldots,\omega_m\}$, we have, by (*) $\omega^2 \neq 0$. Contradiction. □

<u>Proposition 3</u> Let k be algebraically closed and A be a clean k-algebra satisfying the structural property (*) of Proposition 2.

Then A is of minimal rank.

<u>Sketch of the proof</u> Let k algebraically closed and A clean, satisfying (*), as stated in Proposition 2.

Let $\bar{A} := A/\operatorname{rad} A$, $n := \dim_k A$, $p := \dim_k \bar{A}$.

We choose a subalgebra S of A such that $A = S \oplus \operatorname{rad} A$.

Since $S \cong \bar{A} \cong k^p$, we can choose mutually orthogonal idempotents $b_1,\ldots,b_p \in S$ forming a base of S.

Next, note that the right multiplication of A induces a right $\bar{A}$-module structure on L.

Analogously, we consider R as a left $\bar{A}$-module.

Since $\bar{A} \cong k^p$ is semisimple, L and R are semisimple $\bar{A}$-modules too. Therefore, there exist k-bases $\ell_1,\dots,\ell_s$ of L and $r_1,\dots,r_s$ of R such that

(34) $\quad \ell_\sigma A = k\,\ell_\sigma \quad$ for $1 \leq \sigma \leq s \quad$ and $\quad A\, r_\tau = k\, r_\tau \quad$ for $1 \leq \tau \leq t$.

For $1 \leq i \leq m$ let $f_i \in \mathbb{N}$, $f_i \geq 2$, such that $\omega_i^{f_i} \neq 0$, $\omega_i^{f_i+1} = 0$.

Let $\quad q := \dim_k k\omega_1^{f_1} + \dots + k\omega_m^{f_m}$.

Without loss of generality, we may assume that $\omega_1^{f_1},\dots,\omega_q^{f_q}$ are linearly independent.

Let $\quad \delta_i := \begin{cases} 0 & \text{for } 1 \leq i \leq q \\ 1 & \text{for } q < i \leq m \end{cases}$.

Then, by (*),

$$b_1,\dots,b_p,\ell_1,\dots,\ell_s,\omega_1,\dots,\omega_1^{f_1-\delta_1},\ \dots\ ,\omega_m,\dots,\omega_m^{f_m-\delta_m}$$

and

$$b_1,\dots,b_p,r_1,\dots,r_t,\omega_1,\dots,\omega_1^{f_1-\delta_1},\ \dots\ ,\omega_m,\dots,\omega_m^{f_m-\delta_m}$$

generate A over k.

By omitting elements of $\ell_1,\dots,\ell_s$ and $r_1,\dots,r_t$ and renumbering, we may assume that for suitably chosen h, $1 \leq h \leq \min(s,t)$

$$(x_1,\dots,x_n) := (b_1,\dots,b_p,\ell_1,\dots,\ell_h,\omega_1,\dots,\omega_1^{f_1-\delta_1},\dots,\omega_m,\dots,\omega_m^{f_m-\delta_m})$$

and $\quad (y_1,\dots,y_n) := (b_1,\dots,b_p,r_1,\dots,r_h,\omega_1,\dots,\omega_1^{f_1-\delta_1},\dots,\omega_m,\dots,\omega_m^{f_m-\delta_m})$

are bases of A.

Let $X_1,\dots,X_n$, $Y_1,\dots,Y_n$ be indeterminates over k.

To show Proposition 3, it suffices to calculate the product

(35) $\quad (X_1x_1 + \dots + X_nx_n)\,(Y_1y_1 + \dots + Y_ny_n)$,

which we express as a set of bilinear forms belonging to the set of generators of A,

$$b_1,\dots,b_p,\ell_1,\dots,\ell_h,r_1,\dots,r_h,\omega_1,\dots,\omega_1^{f_1-\delta_1},\dots,\omega_m,\dots,\omega_m^{f_m-\delta_m} \in \{x_1,\dots,x_n,y_1,\dots,y_n\}$$

using $2n-p = 2\dim_k A - \#M(A)$ products.

Since $b_1,\ldots,b_p$ are mutually orthogonal idempotents, it follows by (34), that for each σ,τ , $1 \leqq \sigma,\tau \leqq n$, there exists at most one π , $1 \leqq \pi \leqq p$, such that $\ell_\sigma b_\pi \neq 0$, $b_\pi r_\tau \neq 0$ respectively.

Similarly, by (33), to each $1 \leqq i \leqq m$ there exists at most one π, $1 \leqq \pi \leqq p$, such that $\omega_i b_\pi = b_\pi \omega_i \neq 0$.

Therefore, to calculate the bilinear forms belonging to $\ell_1,\ldots,\ell_h$ and $r_1,\ldots,r_h$ one only needs $2h$ products.

In the same way as in [13], Proposition II.3, using for each $i, 1 \leqq i \leqq m$, an interpolation algorithm, one calculates the bilinear forms belonging to

$$b_1,\ldots,b_p, \omega_1,\ldots,\omega_1^{f_1-\delta_1}, \ldots, \omega_m,\ldots,\omega_m^{f_m-\delta_m}$$

by $p + 2(f_1-\delta_1) + \ldots + 2(f_m-\delta_m) = 2(n-h) - p$ products.

So, we can calculate (35) with $2h + 2(n-h) - p = 2n - p$ products. □

Let k be algebraically closed.
Then Corollary 1 (i) is an immediate consequence of Theorem 3 (iii) whereas Corollary 1 (ii) follows from Theorem 3 (ii), taking into account, that the class M_k is closed under homomorphic images.

Finally, we are going to derive Corollary 2 from Theorem 3 in case k algebraically closed.

<u>Lemma 3</u> Let k be algebraically closed, and let T_N be the k-algebra of $N \times N$ upper triangular matrices over k .

Then $R(T_N)$ minimal iff $N = 2$.

For $N = 3$ we have $R(T_3) = 10$, and the trivial 3×3 matrix multiplication algorithm is optimal.

<u>Proof</u> First, note that T_N is clean, the strict upper triangular matrices being its radical.

Let $L_N^* := \{ (a_{ij}) \in T_N \ ; \ a_{ij} = 0 \quad \text{for} \quad i > 1 \}$

and $R_N^* := \{ (a_{ij}) \in T_N \ ; \ a_{ij} = 0 \quad \text{for} \quad j > 1 \}$

One immediately verifies that L_N^* is the left annihilator of $\operatorname{rad} T_N$ and that R_N^* is the right annihilator of $\operatorname{rad} T_N$.

Therefore, L_N^* and R_N^* are two-sided ideals of T_N.

Furthermore, let $L_N := L_N^* \cap \text{rad } T_N = \{x \in \text{rad } T_N \; ; \; x(\text{rad } T_N) = 0\}$ and $R_N := R_N^* \cap \text{rad } T_N = \{y \in \text{rad } T_N \; ; \; (\text{rad } T_N)y = 0\}$.

One immediately verifies for $N \geqq 3$ that $T_N/(L_N^* + R_N^*) \cong T_{N-1}$ holds.
Hence, by Corollary 1 (ii), $R(T_N)$ minimal implies $R(T_{N-1})$ minimal.
So, if $R(T_N)$ is minimal for $N \geqq 4$, then $R(T_3)$ is minimal.

We show by means of Theorem 3 (iii) that this is impossible.

First, notice that $\dim_k \text{rad } T_3 = 3$ and that $\dim_k L_3 = 2$.
Assume, that $R(T_3)$ is minimal.
Then, by Theorem 3, there exists $\omega \in \text{rad } T_3$ with

$$\text{rad } T_3 = L_3 + T_3 \omega T_3 , \quad \text{where} \quad T_3 \omega T_3 = k\omega + k\omega^2 .$$

In particular, $\dim_k T_3 \omega T_3 \leq 2$.
If $\omega^2 = 0$, then $\omega \in L_3$, which is impossible, since $\text{rad } T_3 \not\supseteq L_3$.
Hence $\omega^2 \neq 0$. This implies that ω has the form

$$\omega = \begin{pmatrix} 0 & c_{12} & c_{13} \\ 0 & 0 & c_{23} \\ 0 & 0 & 0 \end{pmatrix} \quad \text{with} \quad c_{12}, c_{23} \in k \setminus \{0\} .$$

Furthermore $$\omega^2 = \begin{pmatrix} 0 & 0 & d_{13} \\ 0 & 0 & 0 \\ 0 & 0 & 0 \end{pmatrix} \quad \text{with} \quad d_{13} \in k \setminus \{0\} .$$

One verifies immediately $$\begin{pmatrix} 1 & 0 & 0 \\ 0 & 0 & 0 \\ 0 & 0 & 0 \end{pmatrix} \omega = \begin{pmatrix} 0 & c_{12} & c_{13} \\ 0 & 0 & 0 \\ 0 & 0 & 0 \end{pmatrix} ,$$

whence $L_3 \subset T_3 \omega T_3$.
Analogously, one shows $R_3 \subset T_3 \omega T_3$.
Since $\text{rad } T_3 = L_3 + R_3$, it follows that $T_3 \omega T_3 = \text{rad } T_3$, whence $\dim_k T_3 \omega T_3 = 3$. Contradiction.

Note that $R(T_2)$ is of minimal rank, the trivial matrix multiplication algorithm being optimal.

This shows the first assertion of Lemma 3.

Since $R(T_3)$ is not minimal, we have $R(T_3) \geq 10$.

But the trivial matrix multiplication algorithm, applied to T_3 , needs 10 products.

Therefore $R(T_3) = 10$, the trivial matrix multiplication algorithm being optimal.

□

Acknowledgment This work was done during the first author's stay at the University of Nice, in Spring 1984. The first author wishes to thank the Mathematical Department of this University for supporting his research.

References

[1] Alder, A. & Strassen, V.: On the algorithmic complexity of associative algebras. Theoret. Comput. Sci. 15 (1981) 201-211.

[2] Atiyah, M.F. & Macdonald, I.G.: Introduction to commutative algebra. Addison - Wesley, Reading MA (1969).

[3] Baur, W.: personal communication (1982).

[4] Behrens, E.-A.: Ring Theory. Pure & Appl. Math. 44; Academic Press, New York (1972).

[5] Bini, D. et al.: $O(n^{2.7799})$ complexity for matrix multiplication. Inform. Proc. Letters 8 (1979) 234-235.

[6] Büchi, W. & Clausen, M.: On a Class of Primary Algebras of Minimal Rank. Linear Algebra and its Appl. 69 (1985) 249-268.

[7] Coppersmith, D. & Winograd, S.: On the asymptotic complexity of matrix multiplication. SIAM J. Comput. 11 (1982) 472-492.

[8] Feig, E.: On systems of bilinear forms whose minimal divisor-free algorithms are all bilinear. J. of Algorithms 2 (1981) 261-281.

[9] Giraldo, L.: Thesis, University of Buenos Aires.

[10] de Groote, H.F.: On varieties of optimal algorithms for the computation of bilinear mappings.
I. The isotropy group of a bilinear mapping.
Theoret. Comput. Sci. 7 (1978) 1-24.

[11] de Groote, H.F.: Characterization of division algebras of minimal rank and the structure of their algorithm varieties.
SIAM J. Comput. 12 (1983) 101-117.

[12] de Groote, H.F.: Lectures on the complexity of bilinear problems.
(to be published) (1982).

[13] de Groote, H.F. & Heintz, J.: Commutative algebras of minimal rank.
Linear Algebra and its Appl. 55 (1983) 37-68.

[14] Lafon, J.C. & Winograd, S.: A lower bound for the multiplicative complexity of the product of two matrices.
(unpublished) (1978).

[15] Schönhage, A.: Partial and total matrix multiplication.
SIAM J. Comput. 10 (1981) 434-455.

[16] Strassen, V.: Gaussian elimination is not optimal.
Numer. Math. 13 (1969) 354-356.

[17] Strassen, V.: Vermeidung von Divisionen.
J. Reine und Angew. Math. 264 (1973) 184-202.

[18] Winograd, S.: On computing the discrete Fourier transform.
Proc. Nat. Acad. Sci. USA 73 (1976) 1005-1006.

[19] Winograd, S.: Some bilinear forms whose multiplicative complexity depends on the field of constants.
Math. System Theory 10 (1977) 169-180.

CONSTRUCTION OF PRIMITIVE IDEMPOTENTS FOR n VARIABLE CODES

A. POLI

AAECC Lab.
Université Paul Sabatier
118 route de Narbonne
31062 Toulouse cédex/France

SUMMARY

We propose an algorithm to construct primitive idempotents in any algebra of the type $A = K[X_1,\ldots,X_n]/(t_1(X_1),\ldots,t_n(X_n))$. Each polynomial t_i has its coefficients in a commutative field K.

INTRODUCTION

Idempotents play a role in various chapters of Algebra : Pierce left decomposition (17), irreducible representation for modules satisfying the Maschke condition (10), Zariski topology (8), partial fraction decomposition (Kung and Tong (7))...
They also play a role in the theory of error-correcting codes : idempotents and quadratic residue codes (4)(18), idempotents and Mattson-Solomon polynomials (18), idempotents and weigth enumerator (18).
Finally, they play a role for polynomial factorization (2),(9).

On the other hand, codes can often be represented in some algebra $A = \mathbb{F}_q[X_1,\ldots,X_n]/(t_1(X_1),\ldots,t_n(X_n))$: cyclic codes (18), two dimensional Fire codes (6), abelian group codes (18), n-variable codes (12).
Idempotents can be useful to obtain the decomposition of A as a direct sum. H.F. De Groote and J. Heinz (5), T. Beth in (1) proved that it is more efficient to work out calculations in each summand of the decomposition of A than in A itself.

So, calculations for constructing codes, or for decoding, can be simplified when using primitive idempotents.

Many constructions where given, for K being a finite field : J.H. Van Lint (1979-15) gave, in fact the Pierce decomposition. P.Camion (1980 (4),1982 (3)) developed a construction in A (A satisfied the Maschke condition).He used,in particular, the construction of the Berlekamp subspace of A and the Pierce technique.He did not give the same calculations for q odd as for q even.
A.Poli (1981 (13),1984 (11)),for group algebra codes,eventually without the Maschke condition,gave a direct computation.
In his thesis M.Ventou (1984 (16)) gave a general result,but he had to perform his calculations in A.

Generalizing (13) we give here a general and direct construction which works even K is of characteristic 0.Our calculations do not need the construction of the Berlekamp subspace of A. Moreover our calculations are done only in some summands of A.

In the sequel we give our notations in Part I.We give all algebraic tools we need in PartII.In Part III we give our algorithm after all necessary proofs.

PART I (NOTATIONS)

K is a commutative field.

For i in $\{1,\ldots,n\}$:

1- Let $t_i(X_i)$ be a polynomial with coefficients in K.

2- Let H_i be the set of all distinct roots μ_i of $t_i(X_i)$.

3- $p_i(X_i)$ is the minimal polynomial of μ_i.Its degree is d_i,and its multiplicity is m_i (in the factorization of t_i).

4- For each $(\mu)=(\mu_1,\ldots,\mu_n)$ in $H_1\times\ldots\times H_n$ we denote by $C(\mu)$ the set of all $(f(\mu_1),\ldots,f(\mu_n))$ with f in the group of K-automorphisms of $K(H_1,\ldots,H_n)$.

5- Let $\bar{t}_i$ be the polynomial,without square root,which admits H_i as set of zeroes. $\bar{A}$ is the algebra $K[X_1,..,X_n]/(\bar{t}_1,..,\bar{t}_n)$.

6- Let $W_i(\mu_1,..\mu_{i-1},X_i)$ be the minimal polynomial of μ_i over $K(\mu_1,..\mu_{i-1})$. We denote by W_i the polynomial $W_i(X_1,..X_i)$ deduced from $W_i(\mu_1,\ldots,X_i)$ substituting X_j to μ_j $(1\leq j\leq i-1)$.

7- Π_i is the polynomial $\Pi_i(X_1,..,X_i)$ which is deduced from $p_i/W_i(\mu_1,..,\mu_{i-1},X_i)$ in the same way.

We remark that in $K[X_1,..,X_n]$the ideal $(p_1,W_2,..,W_n)$ is the (maximal) ideal of all polynomials which admit $(\mu_1,..,\mu_n)$as a zero.

PART II (ALGEBRAIC TOOLS)

In a first step we give an explicit construction for primitive idempotents when K may be of characteristic zero. In a second step we give simplifications when K is of characteristic p (p prime).

The ideal $I=(t_1,..,t_n)$ is an ideal in $K[X_1,..,X_n]$ which is a zero-dimensional one (17). In such a case we know that it is possible to obtain the noetherian decomposition of I, constructively, using Grobner bases (19). Then primitive idempotents of A can be deduced by techniques of linear algebra.

Here we give a direct construction of these idempotents, using a particular family of polynomials defined in (12). Let us recall some useful results we obtain in (12).

Property 1 For A we have the direct sum:

$A=(g_1)\oplus\ldots\oplus(g_N)$ with, for each i in $\{1,..,N\}$:

$g_i=(t_1..t_n/p_1^{m_1}..p_n^{m_n})\ \Pi_2^{m(2)}\ \ldots\ \Pi_n^{m(n)}$; $(m(j)=m_1+..+m_j-j+1; 2\leq j\leq n)$.

For $\bar{A}$ we have, in the same way:

$\bar{A}=(\bar{g}_1)\oplus..\oplus(\bar{g}_N)$ with:

$\bar{g}_i=(\bar{t}_1..\bar{t}_n/p_1..p_n)\ \Pi_2\ldots\ \Pi_n$.

We note that g_i (or $\bar{g}_i$) is construted by using, in particular, a set $p_1,W_2,..,W_n$. This set generates the (maximal) ideal of all polynomials in $K[X_1,...,X_n]$ which admit $(\mu_1,..,\mu_n)$ as a zero. In the sequel we shall write that (g_i) (or $\bar{g}_i$) is associated to $C(\mu)$.

Property 2 We have the two following results:

1-(In A):

$g_i p_1^{i_1}\ W_2^{i_2}\ \ldots W_n^{i_n}=0$ if $i_1+\ldots+i_n\geq m(n)$ holds.

2-(In $\bar{A}$):

$\bar{g}_i p_1^{i_1}\ W_2^{i_2}\ldots W_n^{i_n}=0$ if $i_1+..+i_n\geq 1$ holds.

Now, let g_i be an element of A.

From its construction, we remark that g_i can be denoted by $g_i(m_1,..,m_n)$. Using this notation we have:

Property 3 The nilradical R_i of (g_i) is generated by the polynomials $g_i(m_1-1,m_2,..,m_n)$, $g_i(m_1,m_2-1,..,m_n)$, ..., $g_i(m_1,..,m_{n-1},m_n-1)$.
Each polynomial in $(g_i)\setminus R_i$ is of the type hg_i+r with r in R_i, and

h be invertible modulo $(p_1,W_2,..,W_n)$.
Moreover, each element of R_i is null over $H_1 x..x H_n$. An element of (g_i) is in $(g_i)\setminus R_i$ iff it has not $(\mu_1,..,\mu_n)$ as a zero.

PART III (PROOFS)

Now we construct idempotents in A.

We first suppose that K is of characteristic zero.

Lemma 1 In $(\bar{g}_i)$ let $\bar{h}_i$ be the inverse of $\bar{g}_i$ modulo $(p_1,W_2,..,W_n)$. Then the element $\bar{h}_i\bar{g}_i$ is the primitive idempotent of the ideal (g_i).

For we have:

$\bar{h}_i\bar{g}_i = 1 \bmod (p_1,W_2,..,W_n)$, and (by Proposition 2-2):

$\bar{h}_i\bar{g}_i(\bar{h}_i\bar{g}_i-1)=0$.

Theorem 1 Let M_i be the greatest multiplicity for the roots of t_i.
Let l be equal to $M_1+..+M_n-n+1$.
In A the idempotent e_i, primitive, generating (g_i), is given by:

$e_i=(-1)^{m(n)-1}((\bar{h}_i\bar{g}_i)^l-1)^{m(n)}+1$.

In A let us consider the polynomial $\bar{g}_i$.
By construction it takes the value zero over $H_1 x..x H_n \setminus C(\mu_1,..,\mu_n)$ as we can deduce from Proposition 1. Then (Proposition 2) one has:

$\bar{g}_i=r_1+..+r_n$ (r_j in R_j for $j\neq i$; r_i in $(g_i)\setminus R_i$.

We have:

$\bar{g}_i^{\,l}=r_i^{\,l}$ because $R_j^l = 0$ for all j in $\{1,..,N\}$ (Proposition 3).
Consequently $\bar{g}_i^{\,l}$ is an element of (g_i).

On the other hand, we remark that the congruence:

$\bar{h}_i\bar{g}_i -1=0 \bmod (p_1,W_2,..,W_n)$ implies the next one:

$(\bar{h}_i\bar{g}_i)^l=1 \bmod (p_1,W_2,..,W_n)$. We then deduce:

$((\bar{h}_i\bar{g}_i)^l-1)^{m(n)} =0 \bmod (p_1^{m_1},W_2^{m_2},..,W_n^{m_n})$, which is equivalent to:

$ag_i-1=0 \bmod (p_1^{m_1},..,W_n^{m_n})$ because $\bar{g}_i^{\,l}$ is in (g_i).

Then, by Proposition 2-1, we have:

$ag_i(ag_i-1)=0$.

Example $A=\mathbb{R}[X,Y]/((X^2+1)^2,Y^2+Y+1)$

We have:

$H_1=\{i,-i\}$; $H_2=\{-\sqrt{3}/2\ i-1/2,\ \sqrt{3}/2\ i-1/2\}$.

$H_1 \times H_2 = C(i,-\sqrt{3}/2\ i-1/2) \cup C(i,\sqrt{3}/2\ i-1/2)$.

The polynomial $\bar{g}_1$ associated to $C(i,-\sqrt{3}/2\ i-1/2)$ is the polynomial $Y-\sqrt{3}/2\ X+1/2$. Moreover :

$\bar{g}_1=-\sqrt{3}\ X \bmod (X^2+1, Y+\sqrt{3}/2\ X+1/2)$, and :

$\bar{e}_1=(X/\sqrt{3})(Y-\sqrt{3}/2\ X+1/2)=XY/\sqrt{3}+X/2\sqrt{3}+1/2$.

Furthemore we have:

$l=2$ and $m(2)=2$.

$e_1=(-1)(\bar{e}_1^2-1)^2+1=1/2+(1+2Y)(3+X^2)X/4\ 3$.

From now, K is a finite field $\mathbb{F}_q$ ($q=p^r$, p a prime).

In fact we shall only construct some primitive idempotents in $\bar{A}$ and, then, we shall deduce all others. Finally we shall "lift" to A.

We know by Lemma 1 that $\bar{h}_i\bar{g}_i$ is the primitive idempotent in the ideal $(\bar{g}_i)$.

Let us recall that $(\bar{g}_i)$ is associated to $C(\mu_1,..,\mu_n)$.
Let us recall also that μ_i is a root of $p_i(X_i)$ $(1\le i\le n)$.

We now consider the set Q of all classes $C(\mu'_1,..,\mu'_n)$ such that μ'_i is some root of $p_i(X_i)$ $(1\le i\le n)$.
Let (a,b) (respectively $[a,b]$) denote the GCD (resp. the LCM) of a and b.

<u>Lemma 2</u> The number N of ideals (g_i) (or of ideals $(\bar{g}_i)$) is given by the formula:

$N=\sum d_1..d_n/[d_1,..,d_n]$, where the sum is extended to all possible n-tuples $(p_1,p_2,..,p_n)$.

Clearly N is therefore the number of all classes $C(\mu)$. For a particular choice of $(p_1,..,p_n)$ each class $C(\mu)$ has cardinal equal to $[d_1,..,d_n]$. Then the set Q has cardinal equal to $d_1..d_n/[d_1,..,d_n]$

<u>Lemma 3</u> (14) A set of leaders for classes of Q is:

$$(\mu_1,\mu_2^{q^{l_2}},..,\mu_n^{q^{l_n}});\ 0\le l_i<([d_1,..,d_{i-1}],d_i);\ 2\le i\le n\ .$$

Indeed:

1- The number of such leaders is equal to $d_1..d_n/[d_1,..,d_n]$ which is the cardinal of Q.

2- Two distinct leaders cannot be in the same class.

For let us suppose that we have:

$(\mu_1,\mu_2^{q^{l_2}},\ldots,\mu_n^{q^{l_n}}) \in C(\mu_1,\mu_2^{q^{s_2}},\ldots,\mu_n^{q^{s_n}})$ with the following inequalities: $(0\leq l_i, s_i < ([d_1,\ldots,d_{i-1}],d_i);\ 2\leq i\leq n)$.

From the definition of classes we deduce that there exists t such that the following equalities hold:

$$\mu_1=\mu_i^{q^t}\ ;\ \mu_i^{q^{s_i+t-l_i}}\quad (2\leq i\leq n)\ .$$

The element d_1 divides t. For each i $(2\leq i)$ s_i-l_i is a multiple of $([d_1,\ldots,d_{i-1}],d_i)$. By hypothesis s_i is then equal to l_i. Then $[d_1,\ldots,d_i]$ divides t, and we can iterate for i+1.

<u>Example 2</u> $A=\mathbb{F}_2[X,Y]/((X^2+X+1)^3,(Y^4+Y+1)^2)$.

Here, we have a unique possible choice for (p_1,p_2):

$p_1=X^2+X+1$, $p_2=Y^4+Y+1$.

Then N is equal to 2.4/4 (=2).

Let μ be a primitive element of $\mathbb{F}_{16}$. A set of leaders for Q is $\{(\mu^5,\mu),\ (\mu^5,\mu^2)\}$.

<u>Example 3</u> $A=\mathbb{F}_2[X,Y]/((X^3+X+1)^3,(Y^2+Y+1)^2(Y^6+Y^3+1)^3)$.

Here we have two possible choices for (p_1,p_2):

$p_1=X^3+X+1$, $p_2=Y^2+Y+1$ or Y^6+Y^3+1.

Then N is equal to (3.2)/6 + (3.6)/6 (=4).

Let μ be a primitive element of $\mathbb{F}_{64}$. Let Q be the set of classes associated with $(X^3+X+1,\ Y^6+Y^3+1)$. A set of leaders for Q is $\{(\mu^9,\mu),\ (\mu^9,\mu^2),\ (\mu^9,\mu^4)\}$.

Lemma 1 gives us $\bar{e}_i$, a primitive idempotent in $\bar{A}$.

From $\bar{e}_i$ we deduce now all primitive idempotents of ideals associated with classes of Q.

<u>Property 4</u> Let $\bar{e}_i(X_1,X_2,\ldots,X_n)$ be the primitive idempotent in the ideal $(\bar{g}_i)$ associated with $C(\mu_1,\mu_2,\ldots,\mu_n)$. Then the primitive idempotent of the ideal $(\bar{g}_j)$ wich is associated to $C(\mu_1,\mu_2^{q^{l_2}},\ldots,\mu_n^{q^{l_n}})$ is $\bar{e}_i(X_1,X_2^{q^{d_2-l_2}},\ldots,X_n^{q^{d_n-l_n}})$, for all possible l_i which verifies

$0\leq l_i<([d_1,\ldots,d_{i-1}],d_i)$; $2\leq i\leq n$.

Let φ be the endomorphism of $\bar{A}$ defined by the set of substitutions:

$$X_1 \dashrightarrow X_1\ ;\ X_i \dashrightarrow X_i^{q^{d_i-l_i}}\quad (0\leq l_i<([d_1,\ldots,d_{i-1}],d_i);\ 2\leq i\leq n).$$

Set $u_j = \varphi(\bar{e}_i)$. It is not zero because $u_j(\mu_1,\mu_2^{q^{l_2}},..,\mu_n^{q^{l_n}})$ is equal to 1. As φ is a morphism, u_j is an idempotent.

We examine for which $(\mu_1',..,\mu_n')$ we have $u_j(\mu_1',..,\mu_n')=0$.

By construction of $\bar{e}_i$, the polynomial u_j is a multiple of $\bar{t}_1..\bar{t}_n/p_1..p_n$. Then (Proposition 3) u_j can have a not null component only in ideal $(\bar{g}_k)$ which is associated with some class $C(\mu_1,\mu_2^{q^{s_2}},..,\mu_n^{q^{s_n}})$ with : $0\leq s_i<([d_1,..,d_{i-1}],d_i)$; $2\leq i\leq n$.

We remark that we have:

$$u_j(\mu_1,\mu_2^{q^{s_2}},..,\mu_n^{q^{s_n}})=\bar{e}_i(\mu_1,\mu_2^{q^{d_2-l_2+s_2}},.....,\mu_n^{q^{d_n-l_n+s_n}}).$$

By its construction $\bar{e}_i$ is no zero only for elements of $C(\mu_1,\mu_2,..,\mu_n)$. As for the proof of 2- Lemma 3 , we can deduce that u_j is not zero only for elements of $C(\mu_1,\mu_2^{q^{l_2}},..,\mu_n^{q^{l_n}})$.

Then u_j is a primitive idempotent.

Now we develop the previous examples.

Example 4 The ideal $(\bar{g}_1)$ which is associated to (μ^5,μ) is defined by: $\bar{g}_1= Y^2+Y+X+1$

$\bar{g}_1=1 \bmod (X^2+X+1,\ Y^2+Y+X)$

$\bar{e}_1= Y^2+Y+X+1$

We deduce now $\bar{e}_2$ from $\bar{e}_1$:

$\bar{e}_2=\bar{e}_1(X,Y^8)= Y^{16}+Y^8+X+1= Y^2+Y+X$

Example 5 The ideal $(\bar{g}_1)$ which is associated to (μ^9,μ) is defined by: $\bar{g}_1=(Y^2+Y+1)(Y^2+YX+1)(Y^2+Y(X^2+X)+1)$

$\bar{g}_1=YX+X+1 \bmod (X^3+X+1,Y^2+YX+1)$

$\bar{h}_1=Y(X^2+X)+X$

$\bar{e}_1=Y^7(X^2+X)+Y^6(X+1)+Y^5X^2+Y^4X^2+Y^3(X+1)+Y^2(X^2+X)+X.$

We deduce $\bar{e}_2$ from $\bar{e}_1$:

$\bar{e}_2=\bar{e}_1(X,Y^{32})$

The result can be obtain by a linear algebra technique.

$\bar{e}_2=Y^7X+Y^6(X^2+1)+Y^5(X^2+X)+Y^4(X^2+X)+Y^3(X^2+1)+Y^2X+X^2.$

Also we have:

$\bar{e}_3=\bar{e}_1(X,Y^{16})$

$\bar{e}_3= Y^7X^2+Y^6(X^2+X+1)+Y^5X+Y^4X+Y^3(X^2+X+1)+Y^2X^2+X^2+X.$

Let us give a more complicated example

Example 6 $A=\mathbb{F}_3[X,Y,Z]/(X^{15}-1,Y^{21}-1,Z^{24}-1)$.

$\bar{A}=\mathbb{F}_3[X,Y,Z]/(X^5-1,Y^7-1,Z^8-1)$. We have:

$X^5-1=(X-1)(X^4+X^3+X^2+X+1)$

$Y^7-1=(Y-1)(Y^6+Y^5+Y^4+Y^3+Y^2+Y+1)$

$Z^8-1=(Z-1)(Z+1)(Z^2+1)(Z^2+2Z+2)(Z^2+Z+2)$

We consider the 3-uple $(X^4+X^3+..+1,Y^6+Y^5+..+1,Z^2+1)$. Using the Berlekamp'subspace of the algebra $\mathbb{F}_3[X,Y]/(X^4+X^3+..+1,Y^6+..+1)$ one can deduce a polynomial $(W_2=)\ Y^3+Y^2(X^3+X^2+1)+Y(X^3+X^2)-1$. Also using the Berlekamp'subspace of $\mathbb{F}_3[X,Y,Z]/(X^4+..+1,W_2,Z^2+1)$ one deduces a polynomial $(W_3=)\ X^3+X^2+Z-1$.

Let (μ_1,μ_2,μ_3) be the generic zero of $(X^4+X^3+..+1,W_2,W_3)$. The polynomial $\bar{e}_1$ associated to $C(\mu_1,\mu_2,\mu_3)$ is given by:

$$\begin{aligned}\bar{e}_1=(1+Z^4)\ & a(Y+Y^2)+bY^3+aY^4+b(Y^5+Y^6)+\\ & +Z(2b(Y+Y^2)+aY^3+2bY^4+a(Y^5+Y^6))+\\ & +2Z^2(a(Y+Y^2)+bY^3+aY^4+a(Y^5+Y^6))+\\ & +Z^3(b(Y+Y^2)+2aY^3+bY^4+2a(Y^5+Y^6))\ ,\end{aligned}$$

where $a=1+X+X^4$, and $b=1+X^2+X^3$ hold.

With the same notations as in the Lemma 3 we have:

$d_1=4;\ d_2=6;\ d_3=2;\ 0\le l_2,l_3<2.$

A set of leaders for classes of Q is then:

$\{(\mu_1,\mu_2,\mu_3),\ (\mu_1,\mu_2,\mu_3^3),\ (\mu_1,\mu_2^3,\mu_3),\ (\mu_1,\mu_2^3,\mu_3^3)\}$.

From Proposition 4 we deduce:

$\bar{e}_2=\bar{e}_1(X,Y,Z^3)$

$\bar{e}_3=\bar{e}_1(X,Y^{243},Z)=\bar{e}_1(X,Y^5,Z)$

$\bar{e}_4=\bar{e}_1(X,Y^{243},Z^3)=\bar{e}_1(X,Y^5,Z^3)$.

Let us note that calculations are very easy because A is a group algebra.Note also that all these calculations are available else A is not a group algebra.

Let $(p_1,p_2,..,p_n)$ be a chosen n-tuple. Let Q be the set of all classes associated to this n-tuple (as it is said just before the Lemma 2).

Let j_k be the smallest integer such that $q^{j_k}\ge M_k$ holds for each k such that $(1\le k\le n)$.

Theorem 2 (Lifting to A)

Let $\bar{e}_1(X_1,..,X_n)$ be the primitive idempotent, in $\bar{A}$, which is associated to a class of Q. Then the set of all primitive idempotents, in A, associated to classes of Q is the set of polynomials of the type:

$$\bar{e}_1(X_1^{q^{j_1}}, X_2^{q^{d_2-l_2+j_2}}, \ldots\ldots, X_n^{q^{d_n-l_n+j_n}}) ; 0\le l_i<([d_1,..,d_{i-1}],d_i) ; 2\le i\le n).$$

The mapping h, from $\bar{A}$ to A, which maps X_i on $X_i^{q^{j_i}}$ $(1\le i\le n)$ is a morphism. Then an idempotent is mapped on an idempotent. The value of $h(\bar{e}_1)$ for a n-tuple $(\mu_1',\ldots,\mu_n')$ is equal to the value of $\bar{e}_1$ for $(\mu_1'^{q^{j_1}},\ldots,\mu_n'^{q^{j_n}})$. The correspondence $(\mu_1',..,\mu_n') \dashrightarrow (\mu_1'^{q^{j_1}},\ldots,\mu_n'^{q^{j_n}})$ can be considered as a permutation of $H_1\times\ldots\times H_n$. Clearly this permutation respects classes $C(\mu)$. Then $h(\bar{e}_1)$ is zero for the elements of exactly one class $C(\mu)$, which is a class of Q. $h(\bar{e}_1)$ is a primitive idempotent of A.

Let us develop our examples.

Example 7 $M_1=3$; $M_2=2$; $j_1=2$; $j_2=1$.

$$e_1=\bar{e}_1(X^4,Y^{32})=Y^4+Y^2+X^4+1$$
$$e_2=\bar{e}_1(X^4,Y^{16})=Y^4+Y^2+X^4.$$

Example 8 We deal with the same Q as in the previous example 5. We have: $M_1=3$; $M_2=3$; $j_1=2$; $j_2=2$.

All primitive idempotents, in A, which are associated to classes of Q are of the type $\bar{e}_1(X^4,Y^{2^{8-l_2}})$, with l_2 in $\{0,1,2\}$.

To obtain the precise expression of these idempotents we need to construct the matrix M(X) of the exponentiation by 2 in $\mathbb{F}_2[X]/(X^3+X+1)^3$, and the similar matrix M(Y) in $\mathbb{F}_2[Y]/(Y^6+Y^3+1)^3$.

We obtain:

(for $l_2=2$) $e_1=Y^{17}X^4+Y^{16}(X^4+X^8)+Y^{11}X^4+Y^8X^8+Y^5X^4+Y^4X^8+Y^2X^4+1$

(for $l_2=1$) $e_2=Y^{17}(X^8+X^4)+Y^{16}X^8+Y^{11}(X^8+X^4)+Y^8X^4+Y^5(X^8+X^4)+Y^4X^4+Y^2(X^8+X^4)+1$

(for $l_2=0$) $e_3=Y^{17}X^8+Y^{16}X^4+Y^{11}X^8+Y^8(X^8+X^4)+Y^5X^8+Y^4(X^8+X^4)+Y^2X^8+1.$

Example 9 $M_i=3$; $j_i=1$ $(1\le i\le 3)$; $0\le l_2,l_3<2$.

We have:

$$e_1=\bar{e}_1(X^3,Y^{3^6},Z^{3^2})=\bar{e}_1(X^3,Y^{15},Z^9) \quad \text{(A is a group algebra).}$$
$$e_2=\bar{e}_1(X^3,Y^{15},Z^3);\ e_3=\bar{e}_1(X^3,Y^{12},Z^9);\ e_4=\bar{e}_1(X^3,Y^{12},Z^3).$$

In particular we can calculate the primitive idempotent e_1.

Setting $a'=a(X^3)$ and $b'=b(X^3)$ we have:

$$e_1=\Big[a'(Y^{15}+Y^9)+b'Y^3+a'Y^{18}+b'(Y^{12}+Y^6)+ \\ +Z^9(2b'(Y^{15}+Y^9)+a'Y^3+2b'Y^{18}+a'(Y^{12}+Y^6))+ \\ +2Z^{18}(a'(Y^{15}+Y^9)+b'Y^3+a'Y^{18}+a'(Y^{12}+Y^6))+ \\ +Z^3(b'(Y^{15}+Y^9)+2a4Y^3+b'Y^{18}+2a'(Y^{12}+Y^6))\Big](1+Z^{12}).$$

PART III (ALGORITHM)

In the following we present a very rough outline of the corresponding constructive algorithm.

Step 1- Construction of $M(X_i)$ $(1\leq i\leq n)$

Step 2- Factorization of t_i $(1\leq i\leq n)$. We obtain $\bar{t}_i, p_i, d_i, m_i$.

Step 3- Selection of a n-tuple $(p_1,\ldots,p_n)$

Step 4- Using Berlekamp' subspace in some algebras we construct one n-tuple $(p_1,W_2,\ldots,W_n)$, and the polynomials $\Pi_2,\ldots,\Pi_n$.

Step 5- We calculate $\bar{g}_i$.

Step 6- We reduce $\bar{g}_i$ modulo $(p_1,W_2,\ldots,W_n)$.

Step 7- Determination of $\bar{h}_i$.

Step 8- Determination of $\bar{e}_i$.

Step 9- We do the lifting to A. We obtain all primitive idempotents that we can deduce from $\bar{e}_1$.

Step 10- Return to Step 3 if there is some n-tuple $(p_1,\ldots,p_n)$ left to be selected. If there is no such n-tuple we stop.

CONCLUSION

Primitive idempotents are a useful tool in applied algebra. In this paper we propose a complete method to construct all primitive idempotents in some algebra A over a commutative field K which can be of characteristic 0 as well as p (p a prime).

The method we propose requires only calculations in some sub-algebra of A, and needs the use of linear algebra techniques.

A comparison with a method using Grobner bases will be given in another paper.

REFERENCES

(1) T.Beth "Generalizing the Discrete Fourier Transform"
Discrete Math.,vol.56,n°2-3,pp 95-101,1985.

(2) P.Camion "Improving an algorithm for factoring polynomials over a finite field and constructing large irred.pomyn." IEEE Trans. on Inf. Theory,vol. IT29,n°3,May 1985.

(3) P.Camion "Un algorithme de construction des idempotents primitifs d'idéaux d'algèbre sur $\mathbb{F}_q$". Annals of Discrete Math.,vol.12,pp55-63,1982.

(4) P.Camion "Un algorithme de construction des idempotents primitifs d'idéaux sur $\mathbb{F}_q$" C.R.A.S. Paris,t291,série A (1980).

(5) H.F. De Groote,J.Heintz "Commutative algebras of minimal rank" (preprint).

(6) H.Imai "A theory of two dimensional cyclic codes" Inf. and Control 34,pp1-34,1977.

(7) H.T.Kung,D.M.Tong "Fast algorithms for partial fraction decomp." S.I.A.M. J. Comp.,vol.6,n°3,1977.

(8) J.P.Lafon "Algèbre commutative T2" Chez Hermann,1977.

(9) D.Lazard "Algorithmes fondamentaux en Algèbre commutative" Astérisque 38-39,pp131-138,1976.

(10) M.P.Malliavin "Les groupes finis et leurs représentations complexes". Chez Masson,1981.

(11) A.Poli "Important algebraic calculations for n variable polynomial codes" Discrete Math.,vol.56,n°2-3,pp255-265,1985.

(12) A.Poli "Codes dans certaines algèbres modulaires" Thèse d'Etat,Univ.P.Sabatier,Toulouse,F,1978.

(13) A.Poli "Quelques résultats sur les codes polynomiaux à n variables". Revue du CETHEDEC,4ème Trim.,NS 81-2,pp23-33,1981.

(14) C.Rigoni "Construction of n variable codes". Disc. Math.,vol.56,n°2-3,pp 275-281,1985.

(15) J.H.Van Lint "Coding Theory" Springer Verlag (New York),1973.

(16) M.Ventou "Contribution à l'étude des codes polynomiaux". Thèse de spécialite,Univ.P.Sabatier,Toulouse,F,1984.

(17) B.L.Van der Waerden "Modern Algebra" F.Ungar Pub. Co.,New York,1964.

(18) F.J.MacWilliams,N.J.A.Sloane "The theory of Error Corr. codes" North Holland P.Co.,1977.

(19) F.Winkler,B.Buchberger,F.Lichtenberger,H.Rolletschek "An algorithm for constructing canonical bases (Grobner bases) of polynomial ideals". CAMP.Publ.,n°81-10,Sept.1981.

MULTIVARIATE POLYNOMIALS IN CODING THEORY

HIDEKI IMAI

Department of Computer Engineering
Yokohama National University
Yokohama, Japan 240

In coding theory, more exactly, in the theory of error-correcting codes, multivariate polynomials are often used to represent codewords of multi-dimensional codes.

For example, codewords of a product code can be represented by bivariate polynomials.

Using such polynomial representation, Burton and Weldon proved in 1965 that the product of two cyclic codes is a cyclic code if the lengths of the two cyclic codes are relatively prime.

I think this is the first clear result on two-dimensional cyclic codes.

For multi-dimensional cyclic codes, the polynomial representation of codewords is very convenient and essential.

In this presentation, I will give a brief sketch of the theory and applications of multidimensional cyclic codes.

For the sake of simplicity, I will discuss only binary two-dimensional cyclic codes.

I. TWO-DIMENSIONAL CYCLIC CODES

We should first define two-dimensional cyclic codes, which will be abbreviated as TDC codes hereafter.

Definition

The TDC code **C** of area m x n is defined as a set of m x n arrays with symbols from GF(2) satisfying the three conditions written here:

(1) **C** is a linear code.

(2) The array obtained by shifting the columns of each array in **C** cyclically one unit to the right is also in **C**.

(3) The array obtained by shifting the rows of each array in **C** cyclically one unit downwards is also in **C**.

Cyclic shifts of a code-array are shown in Fig. 1.

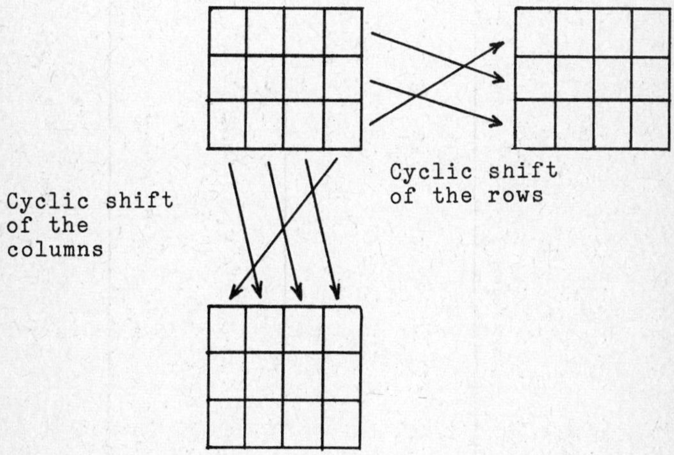

Fig.1

The arrays in **C** are called code-arrays or codewords.

Now, let us see some examples of TDC codes. The first example is a product of two cyclic codes.

[Example 1] Example 1 is a product of a cyclic code {000 101 011 110} with itself. It is a TDC code of area 3 x 3 having the code-arrays shown here:

000	101	110	011
000	011	101	110
000	110	011	101
011	101	011	110
101	000	000	000
110	101	011	110
110	000	000	000
011	101	011	110
101	101	011	110
101	101	011	110
110	101	011	110
011	000	000	000

[Example 2] The TDC code of Example 2 has also 16 code-arrays of area 3 x 3, but it is not a product code.

000	110	011	101
000	110	011	101
000	110	011	101
111	011	101	110
111	100	010	001
000	100	010	001
000	100	010	001
111	011	101	110
111	100	010	001
111	100	010	001
000	100	010	001
111	011	101	110

TDC codes are natural two-dimensional extension of one-dimensional (or conventional) cyclic codes, and they are a special class of Abelian group codes introduced by Berman in 1967 and MacWilliams in 1970.

Consider now why cyclic codes are practically important. One of the reasons is that cyclic codes can be easily encoded by shift registers.

It is owing to this property that cyclic codes have been widely used for error-detection as CRC (cyclic redundancy checking).

For TDC codes to be practically important, they should have the similar property, i.e., their encoding should be easily implemented.

Therefore, let us investigate the encoding of TDC codes. For this purpose, we need some preparations.

Polynomial Representation

As is the case with conventional cyclic codes, it is convenient to use polynomial representation for code-arrays of TDC code. The code-array of equation (1):

$$C = [c_{ij}] = \begin{pmatrix} c_{00} & \cdots\cdots & c_{0,n-1} \\ \vdots & & \vdots \\ c_{m-1,0} & \cdots\cdots & c_{m-1,n-1} \end{pmatrix} \qquad (1)$$

is represented as the bivariate polynomial of equation (2):

$$c(x, y) = \sum_{i=0}^{m-1} \sum_{j=0}^{n-1} c_{ij} x^i y^j. \qquad (2)$$

The polynomial c(x, y) is called a code-polynomial.

[Example 3] The TDC code of area 2 x 2 having the code-arrays of (3):

$$\begin{array}{l} \quad \rightarrow y \\ \downarrow \begin{array}{cccc} 00 & 10 & 01 & 11 \end{array} \\ x \begin{array}{cccc} 00 & 01 & 10 & 11 \end{array} \end{array} \qquad (3)$$

can be expressed as equation (4):

$$C = \{ 0, xy+1, y+x, xy+y+x+1 \}. \tag{4}$$

Using the polynomial representation, we can express each code-polynomial of a TDC code **C** of area m x n as (5):

$$c(x, y) \equiv \sum_{i=1}^{r} a_i(x, y) f_i(x, y) \mod (x^m+1, y^n+1) \tag{5}$$

where f_i are bivariate polynomials of degrees less than m and n with respect to x and y, respectively.

$$\deg_x f_i(x, y) < m$$

$$\deg_y f_i(x, y) < n$$

$\deg_x$ and $\deg_y$ denote the degrees with respect to x and y, respectively.

Clearly, each f_i is itself a code-polynomial of **C**.

Let G be the set of f_1, f_2, ..., f_r, i.e.,

$$G = \{ f_1, f_2, \ldots, f_r \}.$$

Then G specifies the TDC code **C**. We call the set G a basis of **C**, and **C** a TDC code generated by G.

But the basis is not unique for each TDC code.

[Example 4] This example shows some of the bases of the TDC code of Example 2.

$$\{ y^2+y+x^2+x \} \tag{6}$$

$$\{ (x^2+x+1)(y+1), (x+1)(y^2+y+1) \} \tag{7}$$

$$\{ (x^2+x+1)(y+1), y^2+y+x^2+x \} \tag{8}$$

[Example 5] The TDC code of Example 1 has a simple basis shown in (9):

$$\{(x+1)(y+1)\}. \qquad (9)$$

In general, the product code of two cyclic codes generated by the generator polynomials $g_1(x)$ and $g_2(y)$ has a basis of the form of (10):

$$\{ g_1(x)g_2(y) \}. \qquad (10)$$

It is shown that any TDC code of odd area has a basis consisting of a single polynomial.

But it is not necessarily useful to express a TDC code as the set of multiples of the single polynomial.

In general, such expression is rather inconvenient for the implementation of the TDC code.

This is quite different from the case of conventional cyclic codes, where a single polynomial called the generator polynomial plays a central role in encoding and decoding.

Check Positions

In order to construct an encoder for a TDC code, we have to know the check positions, i.e., the positions of parity check bits, or the information positions, i.e., the positions of information bits.

For a conventional cyclic code, these positions are easily determined, because any consecutive bits can be chosen as a set of the check bits if the number of the bits is equal to the degree of the generator polynomial.

But things are different for TDC codes. It is not easy to determine the check positions or the information positions for the TDC code.

In 1974 Imai and Arakaki showed that the check positions can be chosen in the echelon form as in Fig. 2 for any TDC code of odd area.

Their method is based on the zeros of the TDC code, which correspond to the roots of the conventional cyclic code.

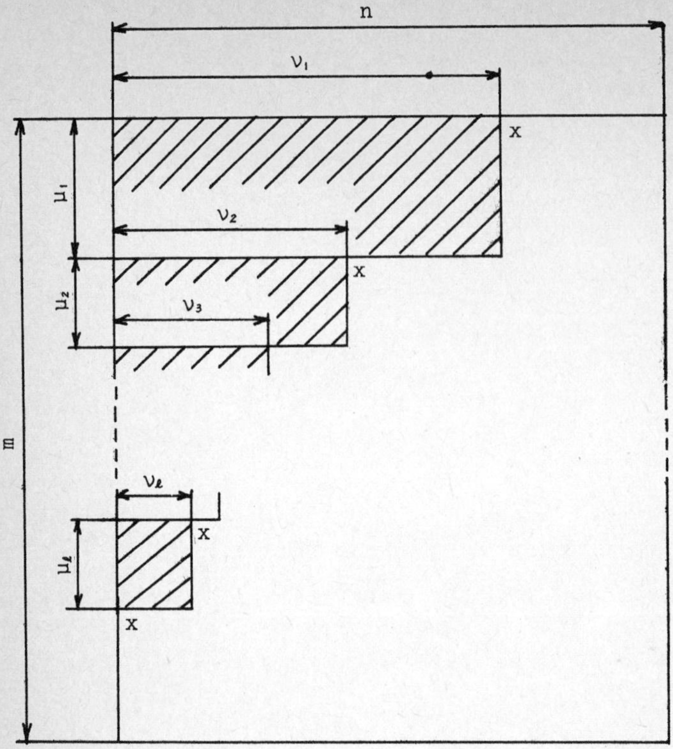

Fig.2

Consider now a TDC code **C** of odd area generated by the basis $\{f_1, f_2, \ldots, f_r\}$.

Let the set of the zeros common to all the polynomials in this basis and x^m+1 and y^n+1 be U.

U is a subset of the direct product of extension fields $GF(2^\mu)$ and $GF(2^\nu)$ which include the m-th and n-th primitive roots of unity, respectively.

Let us define an equivalence relation in U by this:

$$(\xi, \eta) \sim (\theta, \zeta) \tag{11}$$

iff there exists an integer i such that

$$\xi^{2^i} = \theta \tag{12}$$

That is, two zeros are equivalent if and only if their first components are conjugate with respect to GF(2).

According to this equivalence relation, U is partitioned into equivalence classes U_1, U_2, ..., U_ℓ.

Let μ_i be the number of different first components of the zeros in U_i, and we define ν_i by (13):

$$\nu_i = \frac{\text{the number of zeros in } U_i}{\mu_i} \tag{13}$$

It is shown that ν_i is an integer.

Without loss of generality, we assume inequlities (14):

$$\nu_1 \geq \nu_2 \geq \ \cdots \ \geq \nu_\ell \tag{14}$$

Then the echelon form of Fig. 2 can be chosen as the check positions.

[Example 6] The set U for the TDC code generated by the basis of (15):

$$\{ \ y^2+y+x^2+x \ \} \tag{15}$$

is obtained by substituting (16):

$$(x, y) = (\alpha^i, \alpha^j) \quad i = 0,1,2, \ j = 0,1,2 \tag{16}$$

into y^2+y+x^2+x, where α is a root of x^2+x+1 and is a cubic root of unity. Thus we have (17):

$$U = \{ \ (\alpha,\alpha), \ (\alpha^2,\alpha^2), \ (\alpha,\alpha^2), \ (\alpha^2,\alpha), \ (1,1) \ \}. \tag{17}$$

This set is partitioned into U_1 and U_2:

$$U = U_1 \cup U_2 \tag{18}$$

where U_1 and U_2 are given by (19) and (20):

$$U_1 = \{ (\alpha,\alpha), \ (\alpha^2,\alpha^2), \ (\alpha,\alpha^2), \ (\alpha^2,\alpha) \} \tag{19}$$

$$U_2 = \{ (1,1) \} \tag{20}$$

Clearly we have (21) and (22):

$$\mu_1 = 2 , \quad \nu_1 = 2 \tag{21}$$

$$\mu_2 = 1 , \quad \nu_2 = 1. \tag{22}$$

Therefore, the check positions are chosen as the shaded area in Fig. 3.

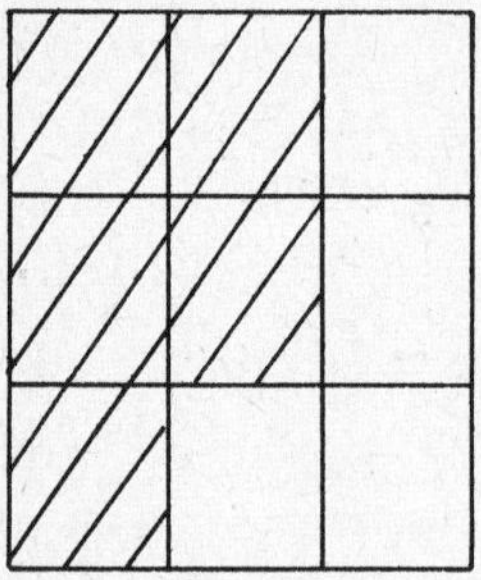

Fig.3

Clearly, the information positions are the positions that are not check positions.

The check positions and the information positions can be chosen as described above.

But they are not unique. Other choices are generally possible.

If the zeros of a TDC code are given, Imai and Arakaki's method for determining the check positions is efficient.

But if a basis is given, the method proposed by Sakata in 1981 is more efficient.

This method is similar to Buchberger's method for constructing reduced Groebner bases, and can be applied not only to the TDC code of odd area but also the TDC code of even area.

I will refer to this method again later. Here let us consider the encoder of a TDC code.

Encoder

An encoding method for the TDC code utilizes a shift register which has storage devices in the form of the check positions, as shown in Fig. 4.

This figure shows the encoder of the TDC code of Example 2.

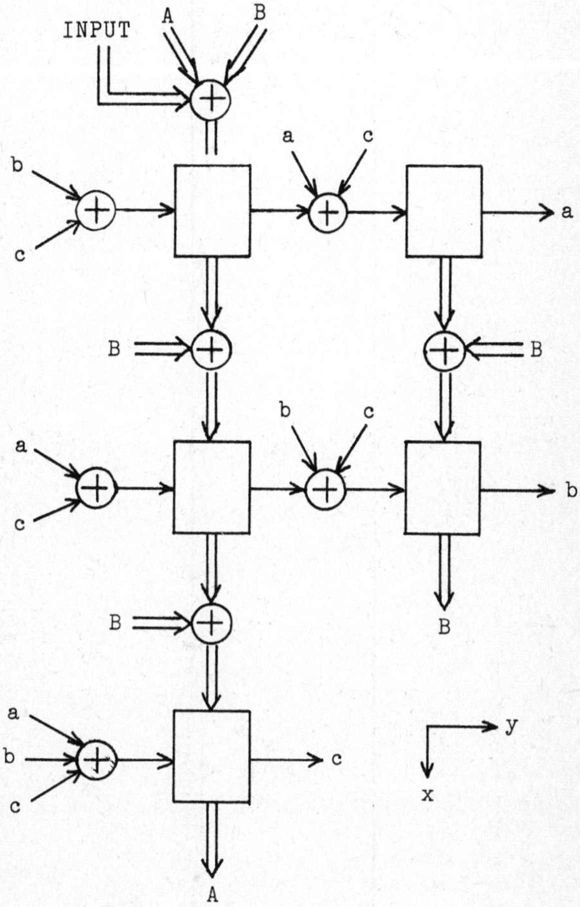

Fig.4

The shift register can be shifted in the direction of x and also in the direction of y.

It has one input line, and the input is added to the input to the storage device at the position (0,0) only when the register is shifted in the direction of x.

In order to encode information bits into a code-array, the storage devices are set to zero initially. Then successive m shifts in the direction of x and one shift in the direction of y are repeated alternatively, and when the register is shifted in the direction of x, each information bits is shifted into the register in the order shown in Fig. 5.

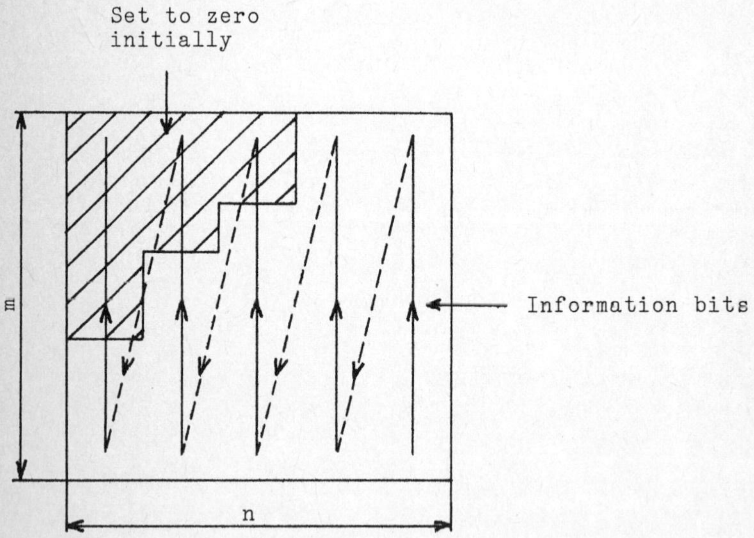

Fig.5

The symbols at the check positions in this figure are set to zero before they are shifted in the register.

The feedback connections of the shift register are determined so that the contents may become the check bits after mn bits of Fig. 5 are read into the register.

The feedback connections can be determined from the zeros or the basis of the TDC code. This will be discussed in more detail later.

There are several other encoding methods for TDC codes. For example the shift register having the storage devices in the form of the information positions can be used for encoding.

Further, if m and n are relatively prime, any TDC code of area m x n is equivalent to a conventional cyclic code of length mn.

Therefore, the encoder for the conventional cyclic code can be used in this case, although it may require more memory than the encoding by the two-dimensional shift register.

The shift register of the type shown in Fig. 4 can also be used as the syndrome calculater and the decoder for two-dimensional burst-error-correction.

Determining the Feedback Connections

Let us now return to the problem of determining the feedback connections of the shift register.

Each feedback connection of Fig. 4 is determined by the polynomials described here:

A	x^3+1
B	$x^2y+xy+y+x^2+x+1$
a	y^2+y+x^2+x
b	$xy^2+xy+x+1$
c	$x^2y+xy+y+x^2+x+1$

The polynomial x^3+1 is used because the register has three rows which is the same as the size of the code in the direction of x.

The rest of the polynomials are code-polynomials.

We should note that these polynomials are of the form (23):

$$x^i y^j + h(x, y) \qquad (23)$$

where (i, j) is the position neighboring the check positions and $h(x, y)$ is the polynomial whose coefficients are confined in the check positions, i.e., $h(x, y)$ can be written as (24):

$$h(x, y) = \sum_{(k,\ell)\in \Pi} h_{k\ell}\, x^k y^\ell \tag{24}$$

where Π denotes the set of the check positions.

Thus in order to determine the feedback connections of the encoder, we should find several code-polynomials of the form (25):

$$c(x, y) = x^i y^j + h(x, y) \tag{25}$$

When the TDC code is of odd area and the zeros are given, then these polynomials can be calculated from the zeros by the method proposed by Imai in 1977.

In general, howovor, it is more efficient and general to construct the reduced Groebner basis, and reduce $x y$ by them, as proposed by Sakata in 1981.

This method will be called Groebner basis method here.

In order to derive the reduced Groebner basis for a TDC code of area $m \times n$ generated by a basis given by (26):

$$\{f_1, f_2, \ldots, f_r\} \tag{26}$$

we generally have to add x^m+1 and y^n+1 to the basis and apply Buchberger's method to the set of (27):

$$\{y^n+1, f_1, f_2, \ldots, f_r, x^m+1\}. \tag{27}$$

The algorithm, being well-known, is omitted here.

If the reduced Groebner basis is obtained, the check positions are easily determined.

Let $G = \{g_1, g_2, \ldots, g_{\ell+1}\}$ be the reduced Groebner basis of the TDC code **C**.

We assume here that the purely lexicographical ordering is used as the total degree ordering in Buchberger's algorithm, although this is not an essential problem.

Then each polynomial in G is written as (28):

$$g_i(x, y) = x^k y^\ell - h(x, y) \tag{28}$$

where (k, ℓ) is the position of the point at the corner marked by x in Fig. 2, and $h(x, y)$ is a polynomial whose coefficients are confined in the shaded area of Fig. 2.

From these (k, ℓ), we can determine the check positions.

For the determination of the feedback connections of the shift register, we use Buchberger's normal form algorithm to $x^i y^j$ and obtain a code-polynomial of the form (30), which determines the feedback connections.

$$x^i y^j + h(x, y) \tag{29}$$

Applications

Let us now turn to applications of TDC codes.

One of the possible applications of TDC codes is two-dimensional burst-error correction, which will be discussed later.

Another application may be found in the study of constructing good random-error-correcting codes. This is further decomposed into two streams.

One is the study to find something like two-dimensional BCH codes.

In 1979 Blahut derived a two-dimensional BCH bound, but it is a very weak bound. Recently Imai and Yamaguchi improved the bound, but it is still weak.

A TDC code which can be constructed algebraically and has better error-correcting capability than the corresponding conventional cyclic code has not been found as yet.

The other study is to find good TDC codes of small area via computer search. This is based on the following facts:

(1) TDC codes of small area are easy to generate systematically.

(2) It is relatively easy to find equivalent codes. Hence we can omit the calculation for the equivalent codes.

(3) The cyclic property of the code simplifies the calculation of the weight of the codewords.

In 1967 Berman showed the minimum distance of the TDC codes of area 3 x 3 and 5 x 5.

In 1981 Sakata computed the spectrum or the weight distribution of the TDC codes of area 2 x 2, 2 x 4, 2 x 6, 2 x 10, 2 x 12, and 4 x 6. Among them some very good codes have been found.

Essentially, TDC codes have more freedom than conventional cyclic codes. So it is natural for them to include better codes.

We may expect that TDC codes better than conventional BCH codes would be found in future.

But it is not easy to find a tight bound for the minimum distance like the BCH bound in one-dimensional case, because of the freedom of TDC codes.

II. TWO-DIMENSIONAL BURST-ERROR-CORRECTING CODES

Two-dimensional burst-error correction may be the most direct application of TDC codes. Let us survey this subject briefly.

There are various information storage deveces that store data on two-dimensional surfaces such as magnetic tapes, magnetic disks, optical disks, and so on.

As the storage density is becoming high, errors in these devices will occur in the form of two-dimensional bursts as shown in Fig. 6.

In order to correct such errors efficiently, we have to introduce two-dimensional structure into the error-correcting code.

Therefore, TDC codes are considered to be appropriate for two-dimensional burst-error correction.

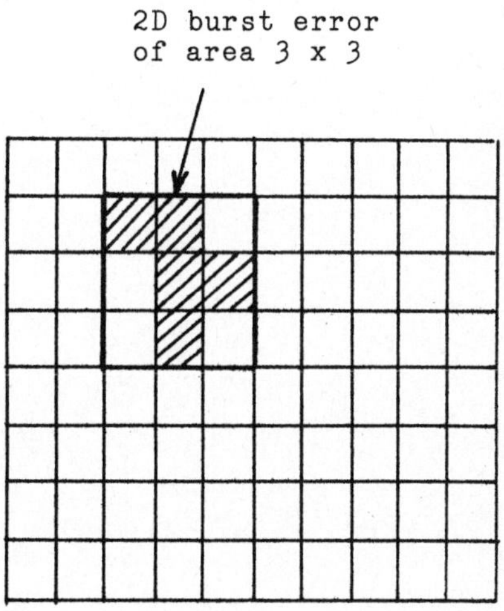

Fig.6

Two-Dimensional Fire Codes

A class of TDC codes for two-dimensional burst-error correction that can be constructed algebraically is the two-dimensional Fire code found by Imai in 1973.

This code will be abbreviated as the TDF code in the following.

In order to define the TDF code, some preparations are necessary.

Let p and q be odd integers and define α and β as the p-th and q-th roots unity, respectively.

Let c and d be integers and define m and n by (30):

$$m = \mathrm{LCM}(c,\ p)\ ,\quad n = \mathrm{LCM}(d,\ q). \tag{30}$$

Then the set of all bivariate polynomials f(x, y) satisfying (31) and (32):

$$\begin{aligned} &\deg_x f(x,\ y) < m \\ &\deg_y f(x,\ y) < n \end{aligned} \tag{31}$$

$$f(\alpha,\ \beta\ ) = 0 \tag{32}$$

is a TDC code of area m x n. We denote this code as $\mathbf{C}_F$.

This TDC code is a very simple code, because it is specified by only one zero.

We now show the reduced Groebner basis of this code.

Let f(x) be the minimal polynomial of α over GF(2). f(x) can be written as (33):

$$f(x) = (x - \alpha)(x - \alpha^2) \cdots\cdots (x - \alpha^{2^{\kappa-1}}) \tag{33}$$

where κ is the minimum positive integer satisfying (34):

$$\alpha^{2^{\kappa}} = \alpha \tag{34}$$

Next let g(x, y) be the polynomial such that the degree with respect to x is less than κ:

$$\deg_x g(x, y) < \kappa \tag{35}$$

and $g(\alpha, y)$ is the minimal polynomial of β over $GF(2^\kappa)$, i.e., the polynomial written as (36):

$$g(\alpha, y) = (y - \beta)(y - \beta^{2^\kappa}) \cdots (y - \beta^{2^{\kappa(\lambda-1)}}) \tag{36}$$

where λ is the minimum positive integer satisfying (37):

$$\beta^{2^{\kappa\lambda}} = \beta \tag{37}$$

Then it is shown that the reduced Groebner basis of the TDC code C_P is given by (38):

$$G = \{f(x), g(x, y)\}. \tag{38}$$

Now we can define the TDF code. The TDF code is a TDC code of area m x n generated by the basis shown in (39):

$$\{x^m+1, y^n+1\}\{f(x), g(x, y)\}$$

$$= \{(x^m+1)f(x), (x^m+1)g(x, y), (y^n+1)f(x), (y^n+1)g(x, y)\}. \tag{39}$$

The meaning of the product of the bases in this expression is obvious and needs no explanation.

The number of the check bits of the TDF code is

$$cd + \kappa\lambda .$$

It is proved that the TDF code is capable of correcting a two-dimensional burst error of area $b_1 \times b_2$ or less, if (40) and (41) are satisfied:

$$c \geq 2b_1-1 , \quad d \geq 2b_2-1 \tag{40}$$

$$\kappa \geq b_1 , \quad \lambda \geq b_2 \tag{41}$$

[Example 8] Let us design a TDF code that can correct a two-dimensional burst error of area 2 x 2 or less.

For this, inequlities (42) should be satisfied:

$$c \geq 3\ , \quad d \geq 3\ , \quad \kappa \geq 2\ , \quad \lambda \geq 2. \tag{42}$$

Therefore, let

$$c = 4\ , \quad d = 3\ , \quad \kappa\lambda = 4 \tag{43}$$

and let γ be a root of primitive polynomial x^2+x+1. Hence γ is a primitive element of $GF(2^4)$.

Define α and β as (44):

$$\alpha = \gamma^5\ , \quad \beta = \gamma^3 \tag{44}$$

Then α is a cubic root of unity and β is a fifth root of unity. We thus have (45):

$$m = LCM(4,\ 3) = 12\ , \quad n = LCM(3,\ 5) = 15. \tag{45}$$

The polynomial f(x) is given by (46):

$$f(x) = (x + \alpha)(x + \alpha^2) = (x + \gamma^5)(x + \gamma^{10})$$

$$= x + (\gamma^5 + \gamma^{10})x + 1 = x^2 + x + 1. \tag{46}$$

On the other hand $g(\alpha, y)$ is written as (47):

$$g(\alpha,\ y) = (y + \beta)(y + \beta^4) = (y + \gamma^3)(y + \gamma^{12})$$

$$= y^2 + \gamma^{10}y + 1 = y^2 + \alpha^2 y + 1$$

$$= y + (\alpha + 1)y + 1. \tag{47}$$

Therefore g(x, y) is calculated as (48):

$$g(x,\ y) = y^2 + (x + 1)y + 1. \tag{48}$$

Then we have (49):

$$\kappa = 2 \ , \quad \lambda = 2 \tag{49}$$

and hence the TDC code can correct a two-dimensional burst-error of area 2 x 2 or less.

A basis of this TDF code is shown here:

$$\{x^4+1,\ y^3+1\}\{x^2+x+1,\ y^2+(x+1)y+1\}$$

$$=\{x^6+x^5+x^4+x^2+x+1,\ (x^2+x+1)y+x^2+x+1,$$

$$(x^4+1)y^2+(x^5+x^4+x+1)y+x^4+1,\ y^5+(x+1)y^4+y^3+y^2+(x+1)y+1\} \tag{50}$$

The reduced Groebner basis for the TDF code is calculated as (51):

$$\{x^6+x^5+x^4+x^2+x+1,\ x^4y^2+y^2+(x^5+x^4+x+1)y,\ y^3+(x^4+1)y+x^5+x^4+x\} \tag{51}$$

From this we can determine the check positions. And the feedback connections is determined by reducing x y by this basis.

Decoding of the TDF code can also be implemented by the two-dimensional shift register that is used for the encoder.

Other Two-Dimensional Burst-Error-Correcting Codes

As two-dimensional burst-error-correcting codes, TDF codes are not very efficient.

It is shown that the number of the check bits r required to correct a two-dimensional burst error of area b_1 x b_2 or less must be no less than $2b_1b_2$:

$$r \geq 2b_1b_2.$$

But for the TDF code r is nearly 5 times b b :

$$r \simeq 5b_1b_2.$$

More efficient two-dimensional burst-error-correcting codes have been constructed from one-dimensional Reed-Solomon codes by Imai in 1976.

But the implementation is a little more difficult than that for the TDF code.

Another class of two-dimensional burst-error-correcting codes have been proposed by Olcayto and Lesz in 1983.

These codes have simple structure and are easily decodable. Further, they can be designed to correct a two-dimensional burst-error of the form other than rectangle, such as discrete circle shown in Fig. 7.

However, since their code construction method requires to check syndromes for all the correctable error patterns, it may be difficult to construct codes of high error-correcting capability.

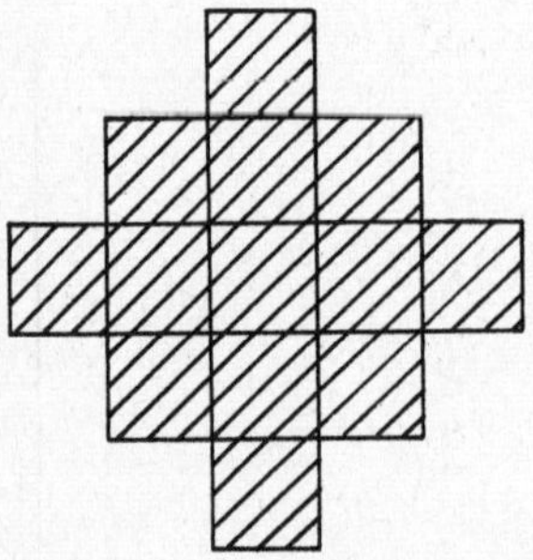

Discrete circle

Fig.7

III. OTHER ASPECTS OF MULTIVARIATE POLYNOMIALS IN CODING THEORY

Other than the study of two-dimensional cyclic codes so far discussed, there are a few areas in coding theory where multi-variate polynomials are utilized.

I only refer to the names of those areas.

Two-Dimensional Linear Recurring Arrays

Two-dimensional linear recurring arrays are periodic arrays generated by linear recurring relations. They are natural two-dimensional extension of linear recurring sequences.

It will be agreed that the most important class of linear recurring sequences are M-sequences (i.e., maximum length shift register sequences).

The two-dimensional extension of M-sequences are called M-arrays, which have several properties analogous to those of M-sequences such as pseudo-randomness.

The general construction method for M-arrays were given by Nomura, Miyakawa, Imai, and Fukuda in 1972.

Their methods are somewhat generalized and refined by Ikai, Kosaka, and Kojima in 1977 and Sakata in 1978.

Another construction method was given by MacWilliams and Sloane in 1976. But I think their method is essentially the same as ours.

Two-dimensional linear recurring arrays are closely related to TDC codes, and have been studied by the same researchres.

The most difficult problem in the study of two-dimensional linear recurring arrays may be to find applications of the theory, as is just the case in the study of TDC codes.

Multivariate Polynomials in Cryptography

Cryptography may not be in the scope of coding theory, but it has at least many relations with the coding theory.

So it is not out of place to mention here the use of multivariate polynomials in cryptography.

In the theory of cryptography, mutivariate polynomials are used to make the expression of some transformation obscure.

Because of the complexity of the appearance of multivariate polynomials, the inverse transformation cannot easily be calculated.

This makes it possible to construct a public key cryptosystem, or an assymmetric cryptosystem.

This idea was found by Matsumoto, Imai, Harashima, and Miyakawa recently.

I think that the practical application of multivaiate polynomials in this area may be more probable than those of TDC codes.

REFERENCES

Arakaki, M., and Imai, H.(1974), "Theory of two-dimensional cyclic codes of even area," (in Japanese), Tech. Report of Inst. Electron. Commun. Eng. Japan, no. AL73-77.

Berman, S.(1967), "On the theory of group codes," Kibernetica, vol. 3, no. 1, pp.31-39.

Berman, S.(1967), "Semisimple cyclic and Abelian codes, II," Kibernetica, vol. 3, no. 3, pp.21-30.

Blahut, R.E.(1979), "Transform techniques for error-control codes," IBM J. Res. Develop. vol.23, pp.299-315.

Buchberger, B.(1976), "A theoretical basis for the reduction of polynomials to canonical forms," ACM SIGSAM Bulletin, vol.39, pp.19-29.

Buchberger, B.(1984), "A survey on the method of Groebner basis for solving problems in connection with systems of multi-variate polynomials," presented at the Second International Symposium on Symbolic and Algebraic Computation by Computers at RIKEN, Japan, pp.7.1-7.15.

Burton, H.O., and Weldon, E.J.,Jr.(1965), "Cyclic product codes," IEEE Trans. Inform. Theory, IT-11, pp.433-439.

Ikai, T., Kosako, H., and Kojima, Y.(1976), "Basic theory of two-dimensional cyclic codes---Generator polynomials and the positions of check symbols," (in Japanese), Trans. Inst. Electron. Commun. Eng. Japan, vol.59-A, pp.311-318.

Ikai, T., Kosako, H., and Kojima, Y.(1977), "Two-dimensional linear recurring arrays---Their formulation by ideals and eventually periodic arrays," (in Japanese), Trans. Inst. Electron. Commun. Eng. Japan, vol.60-A, pp.123-130.

Imai, H.(1972), "Two-dimensional burst-correcting codes," (in Japanese), Trans. Inst. Electron. Commun. Eng. Japan, vol.55-A, pp.385-392.

Imai, H.(1973), "Two-dimensional Fire codes," IEEE Trans. Inform. Theory, vol.IT-19, pp.796-806.

Imai, H.(1976), "A theory of two-dimensional cyclic codes and two-dimensional linear shift registers," (in Japanese), Trans. Inst. Electron. Commun. Eng. Japan, vol.59-A, pp.710-717.

Imai, H.(1977), "A theory of two-dimensional cyclic codes," Inform. and Control, vol.34, pp.1-21.

Imai, H., and Arakaki, M.(1974), "Theory of two-dimensional cyclic codes," (in Japanese), presented at Nat'l Conf. Inst. Electron. Commun. Eng. Japan, no.1415.

MacWilliams, F.J.(1970), "Binary codes which are ideal in the group algebra of an Abelian group," Bell Syst. Tech. J., vol.49, pp.987-1011.

MacWilliams, F.J., and Sloane, N.J.A.(1976), "Pseudo-random sequences and arrays," Proc. IEEE, vol.64, pp.1715-1729.

Matsumoto, T., Imai, H., Harashima, H., and Miyakawa, H.(1984), "An asymmetric bijective cryptosystem with public multivariate polynomials over GF(2)," (in Japanese), Tech. Report of Inst. Electron. Commun. Eng. Japan, no.AL83-77.

Nomura, T., Miyakawa, H., Imai, H., and Fukuda, A.(1971), "A theory of two-dimensional linear recurring arrays," IEEE Trans. Inform. Theory, vol.IT-18, pp.775-785.

Olcayto, E., and Lesz, T.(1983), "Class of linear cyclic block codes for burst errors ocurring in one-, two- and three-dimensional channels," IEE Proc., vol.130, Pt.F, pp.468-475.

Sakata, S.(1978), "General theory of doubly periodic arrays over an arbitrary finite field and its applications," IEEE Trans. Inform. Theory, vol.IT-24, pp.719-730.

Sakata, S.(1981), "On determining the independent point set for doubly periodic arrays and encoding two-dimensional cyclic codes and their duals," IEEE Trans. Inform. Theory, vol.IT-27, pp.556-565.

Sakata, S.(1982), "On the algorithm for determining the normal basis of any multi-variate polynomial ideal," (in Japanese), Proc. of Symp. on Inform. Theory and Its Applications in Japan, pp.7-12.

ENUMERATION OF SELF DUAL 2K CIRCULANT CODES

A. Poli - C. Rigoni

AAECC Lab.
Université Paul Sabatier
118 route de Narbonne
31062 Toulouse cédex/France

SUMMARY

In this paper we give necessary and sufficient conditions for the existence of self dual 2k circulant codes. We specify the number of such codes when these conditions are satisfied.
We prove that all possible 2k circulant codes can be constructed by our method, for every possible code length, and every field $\mathbb{F}_q$.

INTRODUCTION

The study of self dual double circulant codes began in 1969 with a paper of Karlin (3). He remarks that the (24,12,8) extended binary Golay code has a double circulant generator matrix.
Several other codes have such generator matrices : particularly the symmetry codes of V. Pless (7).

In (1) G.F.M Beenker gives a constructive method to obtain double circulant codes over $\mathbb{F}_2$ of $\mathbb{F}_3$, as images of extended cyclic codes over $\mathbb{F}_4$ or $\mathbb{F}_9$.

Also, as circulant codes, several classes of codes where studied by P. Camion, A. Poli, M. Ventou, J. Wolfmann (see (2), (9), (13), (14)).

In this paper we define the 2k circulant codes.
Then we give a method which allows us to contruct all possible self dual 2k circulant codes over any finite field, and for any value of k.

This paper is arranged into three parts.

In the first one we introduce the notations and the algebraic tools used.

In the second one we define 2k circulant codes by one of their generator matrices. Then we specify the necessary and sufficient conditions for this matrix in order to the code be self dual.

In the last part we give the number of self dual 2k circulant codes for every possible code length and every field $\mathbb{F}_q$.

We do not give complete developments for the proofs.

PART I

Some notations. In what follows we shall have to solve an equation in the algebra A defined by :

$$A = \mathbb{F}_q[X_1,\ldots,X_k]/(X_1^{n_1}-1,\ldots,X_k^{n_k}-1)$$

Let :

1) $q = p^r$, p prime
   $n_i = n_i' m_i$, $m_i = p^{e_i}$, $GCD(n_i',p)=1$, $(1 \leq i \leq k)$
   $n = n_1 n_2 \ldots n_k$.

2) Define the polynomial $j(X_1,\ldots,X_k)$ to be the polynomial with all coefficients equal to 1.

For every k-uple $(\mu_1,\ldots,\mu_k)$ of respective roots $\mu_1,\ldots,\mu_k$ of $X_1^{n_1}-1,\ldots,X_k^{n_k}-1$ we give other notations. We will write (μ) instead of $(\mu_1,\ldots,\mu_k)$.

3) Set $C(\mu) = \{(\mu_1^t,\ldots,\mu_k^t),\ t=q^i,\ i \text{ in } \mathbb{N}\}$.

4) $p_i(X_i)$ will denote the irreducible factor of $X_i^{n_i}-1$, of root μ_i.

5) Set $K(\mu) = \mathbb{F}_q(\mu_1,\ldots,\mu_k)$, and $d(\mu) = [K(\mu):K]$.

6) τ will be the involutive automorphism defined by :

$$\tau(R(X_1,\ldots,X_k)) = R(X_1^{-1},\ldots,X_k^{-1}).$$

Now we give some useful properties.

Proposition 1 1) A is decomposed into a direct sum of principal algebras $A(\mu)$. Each of them is generated by its idempotent $e(\mu)$.

2) $\tau(A(\mu)) = A(\mu^{-1})$.

3) $A(\mu)$ is ring isomorphic to $(B(\mu)=)$:

$$K(\mu)\ [Z_1,\dots,Z_k]/(Z_1^{m_1},\dots,Z_k^{m_k})$$

Proof /// These results are consequences of (8) and (9) ///.

Lemma 1. A self reciprocical irreducible polynomial $p(X)$, over $\mathbb{F}_q$, which is different from $X+1$ and $X-1$ has a degree even (say $2t$). We have for each root u of $p(X)$:

$$u^{-1} \doteq u^{q^t}$$

Proof /// Classical ///.

PART II

We define the 2k circulant codes from k-circulant matrices.

Definition of k circulant matrices

The type of a k-circulant matrix is the following :

$$\begin{pmatrix} Q_0 & Q_1 & \cdots & Q_s \\ Q_s & Q_0 & \cdots & Q_{s-1} \\ \cdot & \cdot & \cdot & \cdot \\ Q_1 & Q_2 & \cdots & Q_0 \end{pmatrix}$$

where each matrix Q_i is a $(k-1)$ circulant matrix if k is greater than 1, and is an element of $\mathbb{F}_q$ else.

Definition of 2k circulant codes

A 2k circulant code is a code whose a generator matrix is of the following type :

$$G_0 : \begin{pmatrix} IQ \end{pmatrix} \quad \text{or } G_1 : \begin{pmatrix} a & 0 \dots 0 & c & 1 \dots 1 \\ b & & d & \\ \cdot & I & \cdot & Q \\ \cdot & & \cdot & \\ b & & d & \end{pmatrix}$$

In both cases Q is a k circulant matrix, and a,b,c,d are elements of

$\mathbb{F}_q$.

When the generator matrix is of type G_1 (resp. G_0) then the code is called bordered (rep. pure) code.

Let us give the relation between k circulant matrices and the algebra A.

Lemma 2 The algebra of k circulant matrices is isomorphic to the algebra A.

Proof /// See (11) for example ///.

The polynomial $Q(X_1,\dots,X_k)$ which corresponds to the matrix Q can be recursively defined as a sum of n_k polynomials $Q_i(X_1,\dots,X_{k-1})\ X_k^i$ $(0 \leq i \leq n_k - 1)$. Each polynomial Q_i corresponds to the matrix Q_i.

Property 1 A 2k pure code is self dual iff Q satisfies to :

$$Q\ \tau(Q) = -1.$$

A bordered code is self dual iff Q satisfies to :

$$(I)\ \begin{cases} a^2 + c^2 = -n \\ Q(1,\dots,1) = -ab - cd \\ Q\ \tau(Q) = -1 - (b^2 + d^2)\ j(X_1,\dots,X_k). \end{cases}$$

Proof /// Develop all scalar products involving the two first rows of the generator matrix ///.

In order to simplify the system (I) we use a remark of M. Ventou (12). We construct another generator matrix G_1' which has the same first row as G_1, and whose i^{th} row is the ith row of G_1 plus w times the first row of G_1. Let Q' be the polynomial $Q + w\ j(X_1,\dots,X_k)$.

Proposition 2 The system (I) is equivalent to the following :

(1) $a^2 + c^2 = -n$

(2) $Q'\tau(Q') = -1$ if -1 is a quadratic residue in $\mathbb{F}_q$

$\qquad = -1 + e(1,\dots,1)$ else.

Proof /// The elements $b + wa$, $d + wc$ are deduced from a, c and from the polynomial Q.

To prove proposition 2 it is sufficient to determine a value of w. Choose w as a root of the equation :

$$nx^2 + 2Q(1,\ldots,1)x - (b^2+d^2) = 0 \text{ if } -1 \text{ is q.r. in } \mathbb{F}_q$$
$$= 1/n \quad \text{else} \qquad ///.$$

PART III

In this part we solve equations (1) and (2) of proposition 2.

Proposition 3 The number of pairs which are solutions of (1) is :

$$\begin{cases} q & \text{when } q \text{ is even} \\ q-1 & \text{when } q = 1 \mod 4, \text{ and } k \neq 0 \mod p \\ 2(q-1) & \text{when } q = 1 \mod 4, \text{ and } k = 0 \mod p \\ q+1 & \text{when } q = 3 \mod 4, \text{ and } k \neq 0 \mod p \\ 0 & \text{when } q = 3 \mod 4, \text{ and } k = 0 \text{ mop } p \end{cases}$$

Proof /// See (11) for q odd, direct for q even ///.

Now we solve equation (2).

Using the decomposition of the algebra A into a direct sum one can write : $Q'(X_1,\ldots,X_k) = \Sigma\, r(\mu)$ ($r(\mu)$ is in $A(\mu)$, Q' is a solution of (2)).

We determine the number of elements $r(\mu)$ in each $A(\mu)$.

The proofs are not the same in each case.

1) $A(\mu^{-1}) \neq A(\mu)$

Choose the pair $(r(\mu), r(\mu^{-1}))$ with

$r(\mu)$ is any invertible element in $A(\mu)$,

$r(\mu^{-1}) = -\tau(r(\mu)^{-1})$.

Proposition 4 The number of solutions for the pair $(A(\mu), A(\mu^{-1}))$ is $N(\mu)$ defined by $N(\mu) = (q^{d(\mu)} - 1)\, q^{(m_1 \cdots m_k - 1)\, d(\mu)}$.

Proof /// This number $N(\mu)$ is the number of invertible elements in $A(\mu)$. Recall that $A(\mu)$ is isomorphic to $B(\mu)$. An element of $B(\mu)$ is invertible iff it is a polynomial in $Z_1,\ldots,Z_k$ with a constant coefficient which is not zero ///.

2) $A(\mu^{-1}) = A(\mu)$

In this case we will use the isomorphism between $A(\mu)$ and $B(\mu)$.
Let τ' be the conjugate in $B(\mu)$ of the restriction of τ to $A(\mu)$.
Let $g(Z_1,\dots,Z_k)$ be the corresponding element to $r(\mu)$.
We have to solve, in $B(\mu)$:

$$g(Z_1,\dots,Z_k)\ \tau'(g(Z_1,\dots,Z_k)) = -1.$$

For a sake of simplicity we will write g instead of $g(Z_1,\dots,Z_k)$.

There are two cases to consider

2.1 $d(\mu) > 1$

In this case $d(\mu)$ is even (lemma 1). Set $d(\mu) = 2\ t(\mu)$.
Set $g = \Sigma\ g_{(i)} Z^{(i)}$, with (i) means $(i_1,\dots,i_k)$, and with $Z^{(i)}$ represents the monomial $Z_1^{i_1}\dots\ Z_k^{i_k}$. Elements $g_{(i)}$ are in $K(\mu)$.

Property 2 1) The equation $x\,\tau'(x) = -1$ has $q^{t(\mu)} + 1$ solutions in $K(\mu)$.

2) For every α in $\mathbb{F}_q$ verifying $\alpha\tau'(\alpha) = 1$ there exist $q^{t(\mu)}$ elements x so that $x + \alpha\tau'(x) = w$ for every w verifying $w = \alpha\tau'(w)$.

Proof /// One can use the fact that $\tau'(x)$ is equal to $x^{t(\mu)}$///.

Let us give some more notations concerning the action of τ'.
Set $\tau'(Z_j) = l_j(Z_j)\ Z_j$, and $l^{(i)} = l_1^{i_1}(Z_1)\dots\ l_k^{i_k}(Z_k)$.
One can prove that $l_j(Z_j)\tau'(l_j(Z_j))$ is equal to 1 modulo $Z_j^{m_j-1}$.

Proposition 5 For every polynomial $l^{(i)}$ there exists an element $v_{(i)}$ in $K(\mu)$ such that the set $\{(v_{(i)} + l^{(i)}\tau'(v_{(i)}))\ Z^{(i)}\}$ is a basis F of $B(\mu)$ with every element is left invariant by τ'.

Proof /// Let $l_0^{(i)}$ be the constant coefficient of $l^{(i)}$.
We want that $v_{(i)} + l_0^{(i)}\ \tau'(v_{(i)}) = w$ with w not zero and w be left invariant byt τ'.
Using property 2 we conclude ///.

Corollary 1 There are $N(\mu)$ elements $r(\mu)$, with $N(\mu)$ defined by :

$$N(\mu) = (q^{t(\mu)} + 1)\, q^{(m_1 \dots m_k - 1)t(\mu)}.$$

Proof /// Develop g as a linear combination of the elements of F///

2.2 $d(\mu) = 1$

We have the following property :

Property 3 1) $(\mu) = (\mu_1, \dots, \mu_k)$ with each μ_i equal to 1 or to -1.

2) $\tau'(Z_j)$ is equal to $-Z_j + Z_j^2 - \dots + Z_j^{m_j-1}$ if $\mu_j = 1$ holds, or is equal to $-Z_j - \dots - Z_j^{m_j-1}$ else.

Proof /// Prove that the corresponding element to Z_j, in $A(\mu)$, is $p_j(X_j)\, e(\mu)$, and develop $\tau(p_j(X_j))$ ///.

Now we have to consider two subcases :

2.2.1 q is odd

In this case we can put in evidence a particular basis of $B(\mu)$ which allows to get the number of elements $r(\mu)$. We dont know if ther exists an invariant basis.

Set $(-1)^{(i)} = (-1)^{i_1 + \dots + i_k}$.

Proposition 5 Let F be defined by $F = \{f_{(i)} = (1^{(i)} - (-1)^{(i)})Z^{(i)}\}$.

F is a basis of $B(\mu)$. Each element of F is left invariant by τ' or is sent to its opposite.

Proof /// Direct ///.

From the existence of this basis we deduce the number of elements $r(\mu)$.

Corollary 2 There are $N(\mu)$ elements $r(\mu)$, with $N(\mu)$ defined by :

$$N(\mu) = 2\, q^{(m_1 \dots m_k - 1)/2} \quad \text{if } -1 \text{ is a q.r. in } \mathbb{F}_q,$$
$$= 0 \quad \text{else.}$$

Proof /// Direct ///

2.2.2 q is even

In this case we dont know a particular basis of $B(\mu)$ which directly allows to get the number of elements $r(\mu)$.

Proposition 6 In this case $(\mu) = (1,\dots,1)$ holds.
We have $N(1)$ elements $r(1)$ with $N(1)$ defined by :

$N(1) = 2^D q^{L'}$, with $L' = q_1 \dots q_k - 1 - L$, $L = ((q_1/2)-1)q_2 \dots q_k + 2((q_2/2)-1)q_3 \dots q_k + \dots + 2^{k-1}((q_k/2)-1)$, and where D is the number of q_i's which are strictly greater than 2.

Proof /// The case of an even characteristic is the most difficult. This fact was pointed out by F.J. MacWilliams in her paper (4).

Recall that (μ) is equal to $(1,\dots,1)$ because it is the only possibility (for q_i is even and $A(\mu) = A(\mu^{-1})$ holds).

Prove that $\tau'(Z_j)$ is equal to $Z_j + Z_j^2 + Z_j^3 + \dots Z_j^{m_j-1}$ for each j.

Then use the fact that the matrix of τ' is a Kronecker product $t(m_1,s_1) \otimes \dots \otimes t(m_k,s_k)$ of matrices $t(m_i,s_i)$.
Each such matrix is defined by : $t(m_i,s_i) = (t_{k,l})$ $(0 \leq k,l \leq m_i - 1)$, and with $t_{k,l} = \binom{s_i - j - 1}{k-1}$ ///.

Now we give a theorem which sum up our results.

Let N_1 be the number of invertible elements in algebras $A(\mu)$ which verify : $A(\mu) \neq A(\mu^{-1})$.

Let N_2 be the number of elements $r(\mu)$ in algebras $A(\mu)$ (with $A(\mu) = A(\mu^{-1})$, $d(\mu) \neq 1$) verifying $r(\mu)\,\tau(r(\mu)) = -e(\mu)$.

Let N_3 be the number of elements $r(\mu)$ in algebras $A(\mu)$ (with $A(\mu) = A(\mu^{-1})$, $d(\mu) = 1$, $(\mu) \neq (1)$) satisfying $r(\mu)\tau(r(\mu)) = -e(\mu)$.

Let N_4 be the number of elements x in $A(1)$ such that $x\,\tau(x) = -e(1)$ holds.

Theorem 1) There exist self dual 2k circulant pure codes iff -1 is a q.r. in $\mathbb{F}_q$. The number of such codes is $\sqrt{N_1}.N_2.N_3.N_4$.

2) There exist self dual 2k circulant bordered codes in, and only in, the following cases :

2.1) (q=0,2 mod 4 ; n odd) : the number is $\sqrt{N_1}.N_2.N_4$

2.2) ($q=1 \bmod 4$)

2.2.1) ($n=0 \bmod p,\ p\neq 2$) : $2\ (q-1)\sqrt{N_1}.N_2.N_3.N_4$

2.2.2) ($n\neq 0 \bmod p$) : $(q-1)\sqrt{N_1}.N_2.N_3.N_4$

2.3. ($q=3 \bmod 4,\ n\neq 0 \bmod p,\ n$ odd) : $2\ (q+1)\sqrt{N_1}.N_2$

Proof /// 1) We have to solve $Q'\ \tau(Q')=-1$. See prop. 4, corol. 1, corol. 2, prop. 6.

2.1) As a and c cannot be both zero, n cannot be even. Moreover the number of pairs (a,c) is equal to q (prop. 3). There no exists algebra $A(\mu)$ with $A(\mu) = A(\mu^{-1})$, $d(\mu) = 1$, $(\mu)\neq(1)$.

2.2) The factor $2(q-1)$ comes from prop. 3.

2.3) We cannot have $n=0 \bmod p$ because -1 is not a quadratic residue in $\mathbb{F}_q$.
Moreover, if n is even then there exists at least one algebra $A(\mu)$ such that $(\mu) = (\mu_1,\ldots,\mu_k)$, each μ_i equal to 1 or to -1. Then (corol. 2) there no exists any solution.
When n is odd and not divisible by p, then there no exists algebra $A(\mu)$ with $d(\mu)=1$, $(\mu)\neq(1)$. In A(1) the element r(1) is zero, because it is necessarily a nilpotent, and because (in this case) A(1) is a semi-simple algebra ///.

CONCLUSION

In this paper we give the necessary and sufficient conditions for the existence of self dual 2k circulant codes. For every possible code length we specify the number of such codes.

Our results generalize those obtained by F.J MacWilliams and those obtained by G.F.M. Beenker.

We have not yet obtained significative results concerning automorphisms of the algebra A which would reduce the number of known non equivalent codes. It is a very important open problem.

We thank our referees for their constructive remarks.

REFERENCES

(1) GFM. Beeker
"On double circulant codes"
TH Report 80, WSK 04 July 1980.

(2) P. Camion
"Etude de codes binaires abéliens modulaires autoduaux de petite longueur"
Revue du CETHEDEC, NS79-2.

(3) M.Karlin
"New binary coding results by circulant"
IEEE Trans.Info. Theory,15,(1969),pp797-802.

(4) FJ. MacWilliams
"Orthogonal circulant matrices over finite fields"
Journal of Comb. Th.,10,pp1-17,1971.

(5) FJ. MacWilliams, NJA. Sloane
"The Theory of error correcting codes"
North Holland, Mathematic Library,vol 16(1977)

(6) P.Piret
"Good linear codes of length 27 and 28"
IEEE Trans. on Info. T.,vol. IT-26,n°2,March 1980.

(7) V.Pless
"Symmetry codes over GF(3) and new 5 designs"
Journal of Comb.Theory,12, 1972,pp119-142.

(8) A.Poli
"Codes dans certaines algèbres modulaires"
Thèse de Doctorat d'Etat, UPS Toulouse F,1978.

(9) A.Poli
"Idéaux principaux nilpotents de dimension maximale dans l'algèbre $\mathbb{F}_q G$ d'un groupe abèlien"
Communications in Algebra,Sept.1984,pp 391-401

(10) A.Poli
"Importante algebraic calculations for n variable codes"
Discrete Math.,vol.56,n°2-3,pp255-263.

(11) A.Poli,C.Rigoni
"Codes autoduaux 2k circulant sur $\mathbb{F}$ (q impaire)"
Revue TS,NS,vol.1,n°2,pp211-217

(12) M.Ventou,C.Rigoni
"Self dual double circulant codes"
Disc.Math., vol.56 n°2-3 pp291-298 (1985).

(13) M.Ventou
"Contribution à l'étude des codes correcteurs polynomiaux"
Thèse de spécialité,UPS Toulouse, 1984.

(14) J.Wolfmann
"A new construction of the binary extended Golay code using a group algebra"
Discrete Math.,31,pp 337-338 (1980).

CODES, GROUPS AND INVARIANTS

Thomas Beth
Royal Holloway College
University of London

Introduction

In communcication systems each transmitter/receiver pair usually consists of two parts

- the MODEM, i.e., Modulator and Demodulator

and

- the CODEC, i.e., Coder and Decoder

which are arranged according to the following diagram

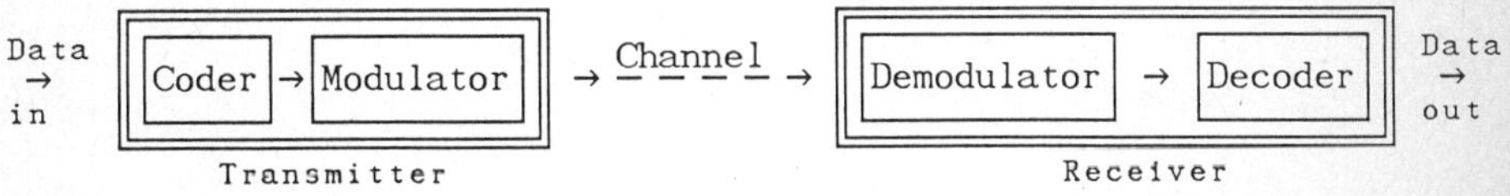

The central aim of today's research in coding theory is the development of techniques which allow the design of most efficient codes with respect to residual - error probabilities and decoding complexity. The recent development of the so-called soft decoding techniques have shown that there is no conceptual difference to be made between the Demodulator and the Decoder cf. Clark/Caine [5].

In the following sections we shall propose a new method by which the features of demodulation and soft-decoding can be considered as a natural application of well understood algebraic methods.

Codes.

For the purpose of this investigation we recall the definition of a code under a special point of view. Following the usual definitions, a code is a specially chosen k-dimensional subspace C of the n-dimensional vector space V_n with a given basis over the field $\mathbb{F}$.

The characteristic parameters of the code are its rate $R = \frac{k}{n}$ and its minimum distance d, cf. MacWilliams-Sloane [6].

The equally important question of decoding complexity in many cases can be related closely to the problem of endowing codes with additional regularity properties. These can be examplified by the action of

Groups

The practical importance of many classes of codes is based on their regularity which is best described by the action of a group G on the vectorspace V_n. To do so we consider codes as G-modules.

Well known examples are

- Reed-Solomon-Codes, where $\mathbb{F} = GF(q)$ and for $n = q - 1$ the cyclic group $G = Z_n$ is acting regularly on the coordinates.

- Quadratic-Residue-Codes, where $\mathbb{F} = GF(2)$ and for n being a prime, $G = ASL(1,n)$ is a group of automorphisms.

In both examples very efficient decoders are constructed using group theoretic properties.

As both codes could simply be considered as cyclic codes, the standard decoding algorithms based on the use of the Discrete Fourier Transform (DFT) of length n, cf. Blahut [4] can be applied. The principle of decoding is usually given by the following diagram

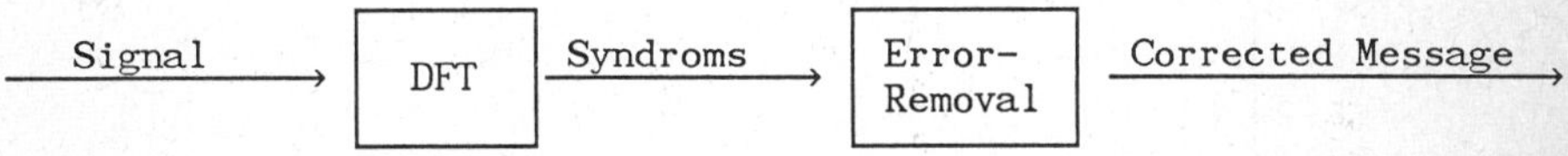

Using the better transitivity properties of the group ASL(1,n) for QRC's, e.g. by applying the so-called error-trapping permutations represents an interesting combinatorial task. We propose a new, more systematic, approach by using

Invariants.

Before giving a rather exact algebraic definition we shall roughly describe the decoding principle proposed in a diagram

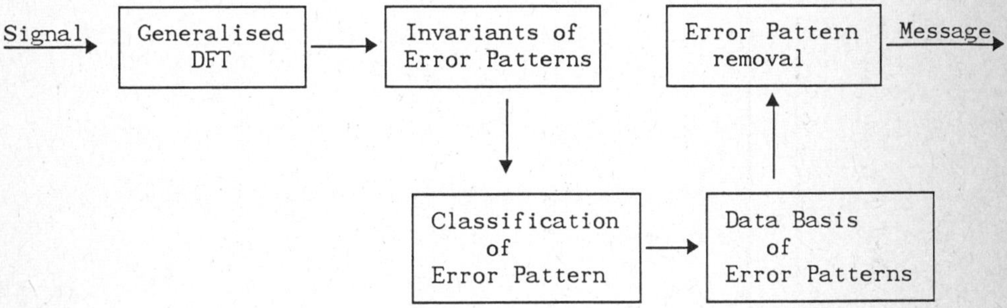

The first step towards the design of this algorithm consists in providing a suitable generalisation of the DFT.

To do so we need a short look at the representations of the group $G = ASL(1,n)$ (n a prime) over a splitting field $\mathbf{K}$ with Maschke condition char $(\mathbf{K}) \nmid |G|$.

The standard tool to obtain an overview for all possible ordinary representations and therefore all G-modules, we consider the group algebra KG. As KG in the present situation is semi-simple we can find all G-modules by considering the simple components A_i of KG. Under the additional assumption that G possesses a suitable series of (sub)-normal-subgroups (e.g., if G is solvable) a fast transform algorithm, the Generalised Fourier Transform.

$$GFT : KG \to \bigoplus_i A_i$$

has been developed (cf. Beth [1], [2]).

In the case of $G = ASL(1,n)$ the series of subgroups is rather trivially given by

$$1 \lhd \mathbb{Z}_n \lhd ASL(1,n)$$

From this series we get a graphic description of the irreducible representations of G, cf. Beth [1].

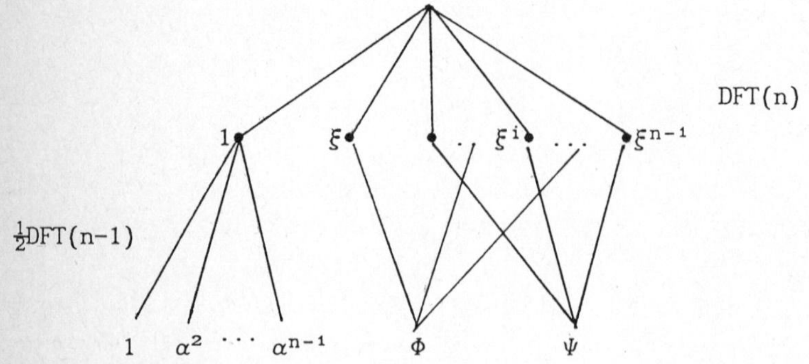

Here ζ denotes a primitive character of $\mathbb{Z}_n$, α a primitive character of $\mathbb{Z}_{n-1}$ and Φ (resp.Ψ) is an irreducible representation of degree $\frac{q-1}{2}$. Indeed, summing up the dimensions of the representations gives

$$\frac{q-1}{2} \cdot 1 + 2 \cdot \left[\frac{q-1}{2} \right]^2 = |G|$$

as required. Before we describe the proposed new decoding method we introduce some necessary notations.

Let $\Phi : G \to M_s(K)$ be a matrix representation of degree s. A polynomial $f \in \mathbf{K}[x_0, \text{—}, x_{s-1}]$ is called an invariant for G iff for all $g \in G$

$$f(x) = f(\underline{x} \cdot \Phi(g)).$$

Application

In order to combine the described algebraic features we note the following

Observation

A code $C < V_n$ considered as a G-module can be described as the intersection of the kernels of certain central idempotents of the simple components $A_{i_1}, \ldots, A_{i_s}$ of the group algebra KG considered as a matrix algebra. In other words:

If $\Phi_{i_1}, \ldots, \Phi_{i_s}$ are the corresponding irreducible representations, then codewords are identified via the following statement:

$$\underline{x} \subset C \iff \forall\, k \in [1 : s]\ \forall\, g \in G: \underline{x} \cdot \Phi_{i_k}(g) = 0.$$

Remark:

The expressions $(\underline{x} \cdot \Phi_{i_k}(g))_{g \in G}$ are the syndroms, which especially in the case of cyclic codes simply reduce to DFT-coefficients, cf. Blahut [4].

Conclusion:

Contains the received message $\underline{u}$ an error pattern, i.e., $\underline{u} = \underline{c} + \underline{e}$, where $\underline{c}$ is the unknown codeword and $\underline{e}$ is a "soft" error pattern (i.e., before demodulation!) then the syndroms

$$((\underline{x} + \underline{e})\Phi_i(g))_g = (\underline{e} \cdot \Phi_i(g))_g$$

deliver a value for the "soft" error pattern.

Observation

Let f be a polynomial invariant for G. Then the evaluation of f at the syndrom $(\underline{e} \cdot \Phi_i(g))_g$ delivers a number $f(x) \in \mathbf{K}$ which for reasons of continuity is stable w.r.t. to the fractional changes that are preserved in a "soft" decoding process.

To apply the tools of invariant theory to this method of error-pattern-classification we have to consider several questions

- How many invariants are available?
- How can we construct them?
- Which among them are good for our purpose?

The answers to the first two questions can be found in coding theory papers where the same question has been investigated for the purpose of determining weight enumerators, cf. Sloane [7], Beth [3], MacWilliams-Sloane [6].

The more important question is the third one which can be more specifically written as

- How many invariant polynomials of low degree are available?
- Which of them give a good separation between classes of error patterns?

The answer to this question has been investigated by I. Zech and the present author [8]. In the following example we demonstrate a typical result.

Example:

Consider the QRC(23) as G-module of the group G = ASL(1,23). Standard calculations show that there are e.g.,

$$\left\{\begin{matrix}1\\8\end{matrix}\right\} \text{ invariants of degree } \left\{\begin{matrix}2\\3\end{matrix}\right\}$$

Let Φ and Ψ denote the two associated irreducible representations of degree 11.

Computations show that for the g classes e_1, ..., e_g of error patterns arising under G.

a) the one invariant f of degree 2 gives the values

i	$f(\underline{e} \cdot \Phi(g))$	$f(\underline{e} \cdot \Psi(g))$
1	11	0
2	21	0
3	30	0
4	30	0
5	30	0
6	30	0
7	30	0
8	30	0
9	30	0

b) a suitable invariant h of the 8 invariant of degree 3 gives the values

i	$h(\underline{e} \cdot \Phi(g))$	$h(\underline{e} \cdot \Psi(g))$
1	11	0
2	19	0
3	21	0
4	21	+5
5	33	-2
6	21	-10
7	21	+10
8	33	+2
9	21	-5

for any g $\in$ G.

This last example shows that the computation of the invariant gives a pointer to the error pattern and therefore the basis to the desired decoding/demodulator-device.

References:

[1] Beth : On the complexity of Group Algebras, to appear in: Theoretical Computer Science

[2] Beth: *Verfahren der Schnellen Fourier Transformation*, Teubner, 1984

[3] Beth: *Codes und Invariantentheorie*, in: Ber. IMMD Erlangen, 11/14, 1978

[4] Blahut: *Theory and Practice of Error Control Codes*, Addison-Wesley, 1983

[5] Clark/Caine: *Error Correcting Codes for Digital Communications*, Plenum Press, 1982

[6] MacWilliams/ Sloane: *The Theory of Error Correcting Codes*, Addison-Wesley, 1983

[7] Sloane: *Error-Correcting Codes and Invariant Theory*, A.M.S. Monthly, Feb. 1977, 82-107

[8] Zech: Diplomarbeit IMMD, Erlangen, 1984

Thomas Beth
Department of Computer Science
Royal Holloway College
Egham, Surrey TW20 OEX
England

ON A CONJECTURE CONCERNING COVERINGS OF HAMMING SPACE.

G. D. Cohen and A. C. Lobstein,
Ecole Nationale Supérieure des Télécommunications,
46 rue Barrault
75634 Paris Cedex 13, FRANCE

and

N. J. A. Sloane
AT&T Bell Laboratories
Murray Hill, NJ 07974, USA

ABSTRACT

We provide evidence for the following conjecture: a minimal covering of the binary Hamming space F_2^{n+2} by spheres of radius $t+1$ has at most the same cardinality as a minimal covering of F_2^n by spheres of radius t.

1. Introduction

For a vector $x \in F_2^n$, we denote by $B(x,t)$ the Hamming sphere of radius t centered at x, i.e. $B(x,t) = \{y \in F_2^n, d(x,y) \leqslant t\}$, where $d(x,y)$ is the Hamming distance between x and y. We say that C is a t-covering of F_2^n if $C + B(\mathbf{0},t) = F_2^n$, where, for $X, Y \subset F_2^n$, $X + Y = \{x+y, x \in X, y \in Y\}$. Finally, let $K(n,t)$ be the minimal cardinality of such a covering.

If moreover C is a linear subspace of F_2^n (a linear code), we call it a linear t-covering, and denote by $k[n,t]$ the minimal dimension of such a covering.

We shall discuss the following conjectures.

Conjecture 1 (the nonlinear case; see [4], [6]):

$$K(n+2, t+1) \leqslant K(n,t) \text{ for all } t \neq n .$$

Conjecture 2 (the linear case; see [1], §10.5):

$$k[n+2,t+1] \leqslant k[n,t] \text{ for all } t \neq n .$$

2. Nonlinear case

Evidence for Conjecture 1 is supplied by the following proposition, which states that for t fixed, $K(n+2,t+1) \leqslant K(n,t)$ for n sufficiently large.

Proposition 2.1 For all t there exists an $N_0 = N_0(t)$ such that

$$K(n+2,t+1) \leqslant K(n,t) \text{ for } n \geqslant N_0 .$$

Proof. Combining the following two results

$$k[n,1] = n - [\log_2 (n+1)] \text{ for } n \geqslant 1$$

(deduced by using Hamming codes), and

$$k[n_1+n_2, 2] \leqslant k[n_1, 1] + k[n_2, 1]$$

we get

$$k[nt,t] \leqslant tn - t[\log_2 (n+1)] .$$

Let N be any integer, $N = nt + r$, $0 \leqslant r < t$. Then

$$k[N,t] = k[nt+r, t] \leqslant k[nt,t] + r \leqslant tn + r - t[\log_2 (n+1)] .$$

Hence

$$k[N,t] \leqslant N - t\left[\log_2 \left(\left[\frac{N}{t}\right] + 1\right)\right] .$$

Also

$$k[N+2,t+1] \leqslant N + 2 - (t+1)\left[\log_2 \left(\left[\frac{N+2}{t+1}\right] + 1\right)\right] .$$

For t fixed, there exists an N_0 such that for any $N \geqslant N_0$ we have

$$\left(\frac{N+2}{2(t+1)}\right)^{t+1} \geqslant 4\sum_{i=0}^{t}\binom{N}{i}.$$

Hence

$$(t+1)\left(\log_2\left(\frac{N+2}{t+1}\right)-1\right) \geqslant \log_2\left(4\sum_{i=0}^{t}\binom{N}{i}\right),$$

$$(t+1)\left[\log_2\left(\left[\frac{N+2}{t+1}\right]+1\right)\right] \geqslant \log_2\left(4\sum_{i=0}^{t}\binom{N}{i}\right).$$

Thus

$$\frac{2^N}{\sum_{i=0}^{t}\binom{N}{i}} \geqslant 2^{N+2-(t+1)\left[\log_2\left(\left[\frac{N+2}{t+1}\right]+1\right)\right]}$$

But

$$K(N,t) \geqslant \frac{2^N}{\sum_{i=0}^{t}\binom{N}{i}}$$

and

$$\begin{aligned} K(N+2,t+1) &\leqslant 2^{k[N+2,t+1]} \\ &\leqslant 2^{N+2-(t+1)\left[\log_2\left(\left[\frac{N+2}{t+1}\right]+1\right)\right]}. \end{aligned}$$

Thus $K(N,t) \geqslant K(N+2,t+1)$ for $N \geqslant N_0$, as required.

We now study Proposition 2.1 in more detail in the cases $t = 1$ and $t = 2$.

The proof of Proposition 2.1 showed that if

$$(t+1)\left[\log_2\left(\left[\frac{N+2}{t+1}\right]+1\right)\right] \geqslant \log_2\left(4\sum_{i=0}^{t}\binom{N}{i}\right)$$

then $K(n+2,t+1) \leqslant K(n,t)$. For $t = 1$ this condition reads

$$2\left[\log_2\left(\left[\frac{N}{2}\right]+2\right)\right] \geqslant 2+\log_2(1+N)\ .$$

This condition is easily seen to hold for $N \geqslant 28$ and $12 \leqslant N \leqslant 15$. Combining this with the known bounds on $K(N,t)$ from [6], we obtain:

Proposition 2.2 $K(N+2,2) \leqslant K(N,1)$ holds for all $N \geqslant 2$, except perhaps for $N = 9$ and $N = 16$.

At present our best bounds for $N = 9$ and $N = 16$ are:

$$54 \leqslant K(9,1) \leqslant 64,\ \ 32 \leqslant K(11,2) \leqslant 56,$$

$$3933 \leqslant K(16,1) \leqslant 4096,\ \ 1533 \leqslant K(18,2) \leqslant 4096.$$

Similarly for $t = 2$ we obtain (the details may be found in [4]):

Proposition 2.3 $K(N+2,3) \leqslant K(N,2)$ holds for $N = 1,3,4,5,6,7$, $20 \leqslant N \leqslant 28$, $43 \leqslant N \leqslant 44$, $91 \leqslant N \leqslant 127$, $187 \leqslant N \leqslant 361$ and $N \geqslant 379$.

3. Linear case

The minimal integer, R say, for which C is an R-covering is called the *covering radius* of C, and we write $t(C) = R$ (see [1]). Let us first give an asymptotic result, similar to Proposition 2.1.

Proposition 3.1 Conjecture 2 holds for t fixed and n sufficiently large.

Proof. See [1], §8.6.

We need some further definitions, introduced in [3]. For any i between 1 and n, let $C_0^{(i)}$ (resp. $C_1^{(i)}$) be the subset of C with i^{th} component equal to 0 (resp. 1). For x in F_2^n and $\epsilon = 0$ or 1 set

$$f_\epsilon^{(i)}(x) = \min\{d(x,c): c \text{ in } C_\epsilon^{(i)}\}$$

and

$$N^{(i)} = \max_{x \in F_2^n} \{f_0^{(i)}(x) + f_1^{(i)}(x)\}$$

(we assume $C_0^{(i)} \neq \emptyset$ and $C_1^{(i)} \neq \emptyset$). Then C has *norm* N if $\min_i N^{(i)} \leqslant N$. Coordinates i such that $N^{(i)} \leqslant N$ are called *acceptable*. If $N \leqslant 2R+1$, C is said to be *normal*. For x in F_2^n, let x'_i be obtained from x by complementing the ith coordinate. Whenever possible we assume $i = 1$ and omit it. Hence C_0, C_1, f_0, f_1, x' will stand for $C_0^{(1)}, C_1^{(1)}, f_0^{(1)}, f_1^{(1)}, x'_1$ respectively. The importance of normality for our purpose stems from the fact that it leads to an efficient construction for linear coverings, the amalgamated direct sum construction (see [3]). This immediately implies:

Proposition 3.2 If $k[n,t]$ is realized by a normal covering, then Conjecture 2 holds.

The remainder of the paper will be devoted to giving sufficient conditions on n and t for the existence of a normal covering. We denote by $x(i)$ the ith coordinate of an element x in F_2^n. Thus $x'_j(i) \neq x(i)$ if and only if $i = j$.

Proposition 3.3 If C has covering radius R, and for some i either $t(C_0^{(i)})$ or $t(C_1^{(i)})$ is at most $R+2$, then C is normal.

Proof. See Lemma 17 of [6].

The minimal distance of C will be denoted by d. Of course $d \leqslant 2R+1$, with equality only for perfect codes.

Definitions. A coordinate i is *good* if $d(C_0^{(i)}, C_1^{(i)}) = d$. A vector x in F_2^n is *bad for* i if $f_0^{(i)}(x) + f_1^{(i)}(x) \geqslant 2R+2$. There exist at least d good coordinates, and because of linearity, we have the following "parallelism" property. For all good i and all c_1 in $C_1^{(i)}$, there is a c_0 in $C_0^{(i)}$ with $d(c_1,c_0) = d$. We will assume from now on that the first coordinate is good. The next result formalises the intuitive feeling that a bad x cannot be "too close" to C.

Proposition 3.4 If C is not normal, then for all good i and all x bad for i, $d(x,C)$ is at least 2.

Proof. See Lemma 21 of [6].

Corollary 3.5. Any linear t-covering with t at most two is normal.

Proof. See Theorem 22 of [6].

Proposition 3.6 For i good and x bad for i,

$$d + 2\,d(x,C) \geqslant 2R + 2\,.$$

Proof. Suppose that $i = 1$ is good, x is bad for i and $f_1(x) \leqslant f_0(x)$, i.e. $d(x,C) = f_1(x)$. Then $f_0(x) \geqslant 2R+2 - f_1(x)$ and $f_0(x) \leqslant f_1(x)+d$ (from the triangle inequality, since i is good) imply $d + 2f_1(x) \geqslant 2R + 2$.

Corollary 3.7. Any code with minimal distance at most 3 is normal.

Proof. If d is at most 3, then (Prop. 3.6) for any bad x, $d(x,C) = R$. Suppose $f_1(x) \leqslant f_0(x)$. Then $f_1(x) = R$ and $f_0(x) \geqslant R + 2$. But now $f_1(x') = R - 1$ (impossible by Prop. 3.6), or $\min\{f_0(x'), f_1(x')\} = R + 1$ (also impossible).

Definitions. A *unique column* for C is a column occuring exactly once in its generator matrix. If such a column exists, we assume it is the first one, and partition $C = C_0 \cup C_1$ according to it. Let k be the dimension of C. Index the nonzero k-tuples in F_2^k with the integers from 1 to $2^k - 1$. We say that the *signature* $S(C)$ of C is $<n_i>$ $(i = 1,2,\ldots,p)$ if G, a generator matrix for C, has n_i columns with index i (see [1], §6.1.3).

Proposition 3.8 If C contains a unique column (the first say), then

– if $x(1) = 0$, we have $f_0(x) \leqslant (n-1)/2$ and $f_1(x) \leqslant (n+1)/2$

– if $x(1) = 1$, we have $f_0(x) \leqslant (n+1)/2$ and $f_1(x) \leqslant (n-1)/2$

Proof. If the first column is unique, then the projection of C_0 and C_1 on the last $n-1$ coordinates gives two codes C'_0 and C'_1 of strength 1, i.e. containing no zero column. Hence $t(C'_0)$ and $t(C'_1)$ are at most $(n-1)/2$ (cf. [5]).

Corollary 3.9. If C contains a unique column and there is a vector x for which $f_0(x) + f_1(x) = n$ then n is odd.

Suppose now that C does not contain a unique column. Then, for every j, n_j is at least 2. After a permutation of columns, we can write $G = [A\,B]$, where A is a $k \times n(A)$ matrix consisting of column j of G taken $2[n_j/2]$ times, for all j (the "even part" of G), with $n(A) = 2\sum_j [n_j/2]$. The matrix B is composed of the remaining $n(B) = \sum_j (n_j - 2[n_j/2])$ columns (the "odd part" of G). Clearly $n(A) \geqslant 2n(B)$, hence $n(A) \geqslant 2n/3$. Also $t(A) \geqslant n(A)/2$. To see this, take x in $F_2^{n(A)}$, $x = (0101 \cdots 01)$. Then $d(x, A) = n(A)/2$. Since $t(C) \geqslant t(A)$ we obtain:

Proposition 3.10 If $t(C) < n/3$, then C contains a unique column.

Starting from a generator matrix $G(C)$ for C, construct $G(\hat{C})$, a generator matrix for a new code $\hat{C}$, called the *contraction* of C, in the following way: take for columns of $G(\hat{C})$ one copy of every column of type j occuring in $G(C)$. For example if C is a repetition code, $G(C) = [1]$; if C is a Hamming code, $G(\hat{C}) = G(C)$.

Lemma 3.11. $\hat{C}$ has length $p \geqslant k$. Furthermore if $p = k$, then min n_j for $j = 1,2,\ldots,p$ is at least d.

Proof. The first sentence is obvious. If $p = k$, then $\hat{C}$ is a $[p,p]$ code with distance 1, hence $d(C) \leqslant \min n_j$.

Suppose from now on that C is not normal (which implies by the previous results $d \geqslant 4, p \geqslant k$ and $R \geqslant 3$) and does not contain a unique column. There are the following cases, using $k \geqslant 3$ and $N \leqslant n$ (see [3]).

A) $k = 3$. A1) $p = 3$. Then $n_j \geqslant d \geqslant 4$ (Lemma 3.11), hence $n \geqslant 12$.

A2) $p = 4$.	$n = 8 \Rightarrow S(C) = <2,2,2,2> \Rightarrow R \geqslant 4$,	$2R + 2 = 10$: contradiction ($N \leqslant n$).
	$n = 9 \Rightarrow S(C) = <3,2,2,2> \Rightarrow R \geqslant 4$:	contradiction.
	$n = 10 \Rightarrow S(C) = <4,2,2,2> \Rightarrow R \geqslant 5$:	contradiction.
	or $S(C) = <3,3,2,2> \Rightarrow R \geqslant 4$:	see Proposition 3.12.
	$n = 11 \Rightarrow S(C) = <5,2,2,2> \Rightarrow R \geqslant 5$:	contradiction.
	or $S(C) = <4,3,2,2> \Rightarrow R \geqslant 5$:	contradiction.
	or $S(C) = <3,3,3,2> \Rightarrow R \geqslant 4$:	see Proposition 3.12.
A3) $p = 5$.	$n = 10 \Rightarrow S(C) = <2,2,2,2,2> \Rightarrow R \geqslant 5$:	contradiction.
	$n = 11 \Rightarrow S(C) = <3,2,2,2,2>$:	same.

B) $k = 4$. B1) $p = 4 \Rightarrow n_j \geqslant 4, n \geqslant 16$.

B2) $p = 5 \Rightarrow n \geqslant 12$: As in A3).

Proposition 3.12 The codes $<3,3,3,2>$ and $<3,3,2,2>$ with dimension 3 and distance at least 4 are unique (up to isomorphism), have $R = 4$ and are normal.

Proof. Consider first $n = 10$. $\hat{C}$ is a $[4,3]$ code with $d(\hat{C}) = 2$ (if $d(\hat{C}) = 1$, then $d(C) \leqslant 3$). So $\hat{C}$ is unique and consists of the even weight vectors of length 4. Similarly for $n = 11$. This gives for C (resp. C') of length 10 (resp. 11, duplicating column 10).

000	000	00	00	0
111	111	11	11	1
111	111	00	00	0
111	000	11	00	0
111	000	00	11	1
000	111	11	00	0
000	111	00	11	1
000	000	11	11	1

Both C and C' have $R = 4$. For C, $d(C) = 4$ and the last 4 coordinates of C are good. By Proposition 3.6, for a bad x, $d + 2d(x,C) \geqslant 2R + 2$, hence $d(x,C) \geqslant 3$. It is enough (see [3]) to test x with weight at most 4 to check that the good coordinates are acceptable. For C', $d(C') = 5$, and every coordinate is good and acceptable.

Summarizing:

Proposition 3.13 Every code with $n \leqslant 11$ and not containing a unique column is normal.

Suppose now C contains a unique column. We denote by $t[n,k]$ the least covering radius of any $[n,k]$ code (cf. [1]).

Proposition 3.14. If C contains a unique column and has dimension 3, then C is normal.

Proof. If n is even, $t[n,3] = n/2 - 1$, hence $2t[n,3] + 2 = n$ ([1]). If C is not normal, then $N = n$, contradicting Corollary 3.9. If n is odd, $t[n,3] = (n-1)/2 - 1$, hence $2t[n,3] + 2 = n - 1$. Take $i = 1$ and x with $x(1) = 0$ bad. Then by Proposition 3.8, $f_0(x) \leqslant (n-1)/2$ and $f_1(x) \leqslant (n+1)/2$. But $\min\{f_0(x), f_1(x)\} \leqslant R \leqslant (n-1)/2 - 1$; if $f_1(x) \leqslant f_0(x)$, then $f_1(x) + f_0(x) \leqslant (n-1)/2 - 1 + (n-1)/2 = n - 2$, a contradiction. So we must have $f_0(x) < f_1(x)$, and $f_0(x) \leqslant (n-1)/2 - 1$, $f_1(x) \geqslant (n+1)/2$, which imply $f_0(x) = (n-1)/2 - 1$, $f_1(x) = (n+1)/2$, since $f_0(x) + f_1(x) \geqslant n - 1$. But now for x' we have $f_0(x') = (n-1)/2$ and $f_1(x') = (n-1)/2$, contradiction.

From Proposition 3.8 $f_0(x) + f_1(x) \leqslant R + (n+1)/2$ for all x. In fact we can say a little more. Suppose equality, and $x(1) = 0$. Then $f_0(x) = R$ and $f_1(x) = \left[\frac{n+1}{2}\right]$. And for x',

$f_0(x') = R + 1$ and $f_1(x') = \left[\frac{n-1}{2}\right]$, a contradiction. This proves:

Proposition 3.15 If C is not normal and has a unique column then $f_0(x) + f_1(x) \leqslant R + (n-1)/2$ hold for all x.

Combining this with $2R + 2 \leqslant f_0(x) + f_1(x)$, we obtain:

Proposition 3.16 If C is not normal and contains a unique column, then $n \geqslant 2R + 5$.

The smallest unsettled case is $n = 11$, $R = 3$, $k = 4$, $d = 4$.

Proposition 3.17 If C contains a unique column, and has dimension 4, then C is normal for even n.

Proof. Analogous to Proposition 3.14: $R \geqslant t[n,4] = n/2 - 2$. So $2R + 2 \geqslant n - 2$. By Proposition 3.15 $f_0(x) + f_1(x) \leqslant R + (n-1)/2 < 2R + 2$. Q.E.D.

For odd n, the only possibility with $k = 4$ is $d \geqslant 4$, $R = (n-5)/2 = t[n,4]$.

Conclusion. In this paper we establish Conjecture 1 when $t \leqslant 2$, except possibly for a finite number of cases. Combining Propositions 3.2-3.12, we see that Conjecture 2 holds if one of the following hypotheses is true: $n \leqslant 10$, $t \leqslant 2$, or $d \leqslant 3$. The study of the nonlinear case is continued in [6], and the linear case in [7]-[9].

Note added in proof. Since this paper was written it has been established that all binary linear codes with $n \leqslant 14$, or $k \leqslant 5$, or $d \leqslant 5$ are normal [7], [8], [9]. Conjecture 2 has an analog in terms of $t[n,k]$, namely

$$t[n+2,k] \leqslant t[n,k]+1 .$$

This is now known to be true when n is large with respect to $n-k$ [1] or with respect to k [8]. Furthermore $t[n,k]$ is realized by a normal covering for fixed k and sufficiently large n [8].

References

[1] G. D. Cohen, M. R. Karpovsky, H. F. Mattson, Jr. and J. R. Schatz, *Covering Radius - Survey and Recent Results*, IEEE Trans. Inform. Theory, **IT-31** (1985), 328-343.

[2] A. C. Lobstein, G. D. Cohen and N. J. A. Sloane, *Recouvrements d'espaces de Hamming binaires*, Comptes Rendus Acad. Sci. Paris, Série I, **301** (1985), 135-138.

[3] R. L. Graham and N. J. A. Sloane, *On the Covering Radius of Codes*, IEEE Trans. Inform. Theory, **IT-31** (1985), 385-401.

[4] A. C. Lobstein, *Contributions au codage combinatoire: ordres additifs, rayon de recouvrement*, These de docteur-ingenieur, Paris, 1985.

[5] T. Helleseth, T. Klove and J. Mykkeltveit, *On the Covering Radius of Binary Codes*, IEEE Trans. Inform. Theory, **IT-24** (1978), 627-628.

[6] G. D. Cohen, A. C. Lobstein and N. J. A. Sloane, *Further results on the covering radius of codes*, IEEE Trans. Inform. Theory, (1986), to appear.

[7] N. J. A. Sloane, *A new approach to the covering radius of codes*, J. Combinatorial Theory (1986), to appear.

[8] K. E. Kilby and N. J. A. Sloane, *On the covering radius problem for codes:* (I) *Bounds on normalized covering radius*, SIAM J. Alg. Disc. Methods, to appear.

[9] K. E. Kilby and N. J. A. Sloane, *On the covering radius problem for codes:* (II) *Codes of low dimension; normal and abnormal codes*, SIAM J. Alg. Disc. Methods, to appear.

An Improved Upper Bound on Covering Radius

by

H. F. Mattson, Jr.

School of Computer and Information Science

Syracuse University

Syracuse, N.Y. 13244-1240

Part of this paper was presented at the Fourth Caribbean Conference on Combinatorics and Computing, San Juan, April 1-4, 1985.

Abstract

A simple upper bound on covering radius yields new information on various codes. It leads us to show that the nonlinear codes of Sloane and Whitehead [18] are quasi-perfect. We get some new bounds for the Berlekamp-Gale switching problem [7]. It gives the exact covering radius for some codes of length up to 31 and is within 1 or 2 of the exact value for the even quadratic-residue codes of lengths 41 and 47.

Introduction

The covering radius of a code C is the least integer $t(C)$ such that the set of all spheres of radius $t(C)$ centered at the codewords covers the containing space. We confine this paper to binary codes.

In this paper is a generalization — to its "proper" linear version, I hope — of an earlier bound [13] on $t(C)$ for linear codes C. There is also a nonlinear version. It gives information on the covering radius of a variety of linear and nonlinear codes.

For linear codes C, $t(C)$ is the weight of a coset of greatest weight. For more information, the reader may consult [3].

Notation

By the term n - vector we mean an element of Z_2^n.

We often denote an n-vector x by its *support*, the subset of coordinate-places where the 1's of x are. Thus if x is $(x_1, \ldots, x_n)$ in the usual notation, we might also think of it as

$$x = \{\, i \,;\, i = 1, \ldots, n, x_i = 1\}.$$

Then the weight of x is $|\, x \,|$, the cardinality of its support. A *descendant* of a vector x is any vector $y \subseteq x$.

I_r stands for the $r \times r$ identity matrix, and 0^r, 1^r for r-vectors (either row- or column-, as appropriate) of all 0's or all 1's.

We denote a generator matrix of the linear code C by $g(C)$.

$d(C)$ denotes the minimum distance in the code C.

A codeword is called *acarpous* if none of its descendants (except 0) is a codeword.

If C is any code, and $C, 0$ denotes the code obtained by appending an identically 0 coordinate, then $t(C,0) = 1 + t(C)$. For this reason *we restrict this paper to codes having no coordinates identically* 0.

We will sometimes refer to $t[n, k]$, the least covering radius of any linear $[n, k]$ code [3, 7].

The Main Result

The purpose of this paper is to present and explore the following bound on covering radius.

Let C be an $[n, k]$ code. Let J be any subset of the coordinate-places of C. Consider a generator matrix $g(C)$ of C of the form

$$g(C) = \begin{array}{|c|c|} \multicolumn{1}{c}{J} & \multicolumn{1}{c}{} \\ \hline g(C_J) & A \\ \hline 0 & g(C_0) \\ \hline \end{array}$$

Figure 1.

where we have permuted the columns so that J is at the left. The code C_J is the projection of C onto the coordinate-places J; C_0 is the subcode of C which is 0 on J, shortened by the removal of the coordinates J.

THEOREM 1. $t(C) \leq t(C_J) + t(C_0)$.

Proof. If $x = (x', x'')$ is any n-vector, where x' is a $|\, J \,|$-vector, then there is a vector $a = (a', a'')$ in C such that $|\, a' - x' \,| \leq t(C_J)$. To a we may add the vector $b = (0, b'')$ of C such that $|\, (x'' + a'') - b'' \,| \leq t(C_0)$. Thus x is at distance at most $t(C_J) + t(C_0)$ from the codeword $a + b$. QED.

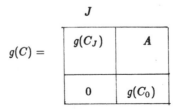

Notice that Theorem 1 remains true if we replace C_0 by any subcode of C_0, though the bound may be worse.

This simple result allows us to find good upper bounds on the covering radius in a number of cases. Before giving examples of such codes we make a few remarks.

REMARK 2. Theorem 1 generalizes the main result of [13], in which J was chosen as an information set of $C^{\perp}$. The bound of Theorem 1 equals the exact value of the covering radius for each of the [10,4,4] codes presented in [13].

REMARK 3. C_J is a $[|J|, j]$ code, and C_0 is an $[n - |J|, k - j]$ code, for some integer j with $0 \le j \le k$. If we let C' be the code of length $n - |J|$ spanned by the rows of A and of $g(C_0)$, then from [3, §3.4], C is a catenation of C_J and C', and

$$t(C_J) + t(C') \le t(C) \le t(C_J) + t(C_0) \le n - k.$$

Karpovsky [11] has observed that if a check matrix of C in the form I_{n-k}, D has a column of $n - k$ 1's, then $t(C) \le \lceil (n-k)/2 \rceil$. This approach can be generalized:

COROLLARY 4. If C has a check matrix equivalent to

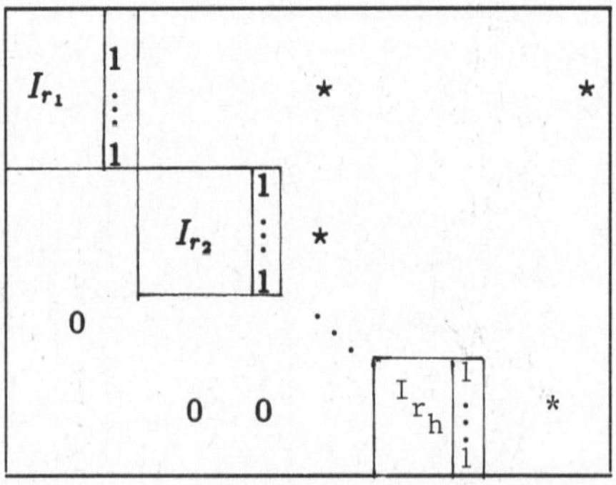

Figure 2.

then

$$t(C) \le \sum_{1 \le i \le h} \lceil r_i/2 \rceil.$$

To prove the Corollary we use induction on h. Take J to be the first $1 + r_1$ coordinate-places; then C_J is the $[n_1, 1, n_1]$ code, where $n_1 = 1 + r_1$. Since $t(C_J) = \lfloor n_1/2 \rfloor$, the result follows. (It is also easy to prove the Corollary directly by arguing on the syndromes.)

Although this Corollary is sometimes useful (see below), it is weaker than the Theorem. For example, the latter easily produces an upper bound of 2 for the Hamming code of length $2^m - 1$, but the Corollary could not give a better bound than $m/2$.

Analogues of the Griesmer Bound [8]

Take J to be the complement of a vector of minimum weight $d(C) = d_1$. Then C_0 is the $[d_1, 1, d_1]$ code of covering radius $\lfloor d_1/2 \rfloor$.

COROLLARY 5. $t(C) \leq \lfloor d_1/2 \rfloor + t(C_J)$, and

$$t(C) \leq \sum_{1 \leq i \leq k} \lfloor d_i/2 \rfloor,$$

where, for $i \geq 2$, d_i is the minimum distance in the subcode of C 0 on all the coordinate-places supporting the previously chosen vectors of weights $d_1, \ldots, d_{i-1}$, respectively.

In other words, we make a Griesmer decomposition of C, and apply Theorem 1. The iterated version is not too useful, because it says only that $t(C) \leq (n-a)/2$, where a is the number of odd d_i's. Still, it offers an amendment to the familiar redundancy bound:

COROLLARY 6. For any linear $[n, k]$ code C having no coordinates identically 0,

$$t(C) \leq min\{n-k, \lfloor n/2 \rfloor\}.$$

The $[2^m - 1, m, 2^{m-1}]$ simplex code has covering radius $\lfloor n/2 \rfloor$ [12, p. 173].

We generalize Corollary 5:

COROLLARY 7. If J is any acarpous codeword of C (in particular, if $|J| < 2d(C)$), then $t(C) \leq \lfloor |J|/2 \rfloor + t(C_{\bar{J}})$.

This Corollary may be helpful because the redundancy of $C_{\bar{J}}$ is $|J| - 1$ less than that of C, enabling one sometimes to calculate $t(C_{\bar{J}})$ by exhaustion.

Now we make a Griesmer decomposition of $C^\perp$, the $[n, n-k, d]$ code orthogonal to C. If J is a minimum-weight vector in $C^\perp$, then in the notation of Theorem 1, C_J is the $[d, d-1, 2]$ code because the all-1 vector J of length d is the only linear dependence on the coordinates in J.

COROLLARY 8. $t(C) \leq 1 + t(C_0)$, where C_0 is the subcode of C of all vectors 0 on the support of a vector of minimum weight in $C^\perp$.

Iteration of this last bound yields only the redundancy bound $t(C) \leq n-k$, if $d(C) \geq 2$.

Let us generalize the approach of Corollary 8. Letting J be any subset of coordinate places, we decompose the check matrix of C as in Figure 1. We have

$g(C^\perp) =$

J	
$g(C_J^\perp)$	0
$*$	$g(C_0^\perp)$

Figure 3.

In symbols

$$(C_0)^\perp = (C^\perp)_{\bar{J}}.$$
$$(C_J)^\perp = (C^\perp)_0 \quad \text{for } \bar{J}$$

COROLLARY 9. If J is any acarpous codeword of $C^{\perp}$, then $t(C) \leq 1 + t(C_0)$. Here C_0 is a $[n- \mid J \mid, k- \mid J \mid +1]$ code with redundancy $n - k - 1$.

Further Results

Now we examine $[n, k, d]$ codes C which have a vector x of weight $2d$. We take $J = \overline{x}$, and $j = \dim C_J$.

THEOREM 10. The shortened subcode C_0 of the $[n, k, d]$ code C consisting of a codeword x of weight $2d$ and its descendant codewords, of type $[2d, k-j, d]$ (or $[2d, 1, 2d]$ if $k - j = 1$), is the $(d/2^{k-j-2})$- fold repetition of the first-order Reed-Muller code R_{k-j-1} of type $[2^{k-j-1}, k - j, \lceil 2^{k-j-2} \rceil]$.

(This barbarous expression is used to show that the limiting case $j = k - 1$, i.e., in which x is acarpous and C_0 is the $[2d, 1, 2d]$ code, fits the same pattern as the general case.)

Proof. To simplify the notation, we prove that any $[2n, k, n]$ code C having 1^{2n} in it is the $n/2^{k-2}$– fold repetition of the $[2^{k-1}, k, 2^{k-2}]$ 1st-order Reed-Muller code. If we "remove" the all-1 vector from C we get a $[2n, k-1, n]$ code C' in which all vectors ($\neq 0$) have weight n. C' is therefore [15, p. 61] the r-fold repetition of the $[2^{k-1} - 1, k - 1, 2^{k-2}]$ simplex code with possibly some columns of 0's appended, say b in number. Equations now for the weight and length are

$$r(2^{k-2} - 1) + b = n$$
$$r(2^{k-1} - 1) + b = 2n,$$

which imply $r2^{k-2} = n$ and $b = r$. QED.

REMARK 11. The "double" of any linear code B of length n has covering radius n. (The "double" of B is the code $2B = \{(x, x); x \in B\}$.)

Proof. The vector $(1^n, 0^n)$ is at distance n from $2B$. In fact, for any x, $(x, \overline{x})$ is at distance n from every vector of $2B$. In any linear code with length N and no coordinates identically 0, every coset has average weight $N/2$. (This last remark furnishes a second simple proof of Corollary 6.)

In the light of Remark 11, it is trivial to apply Theorem 10 when the repetition-factor is even. The covering radius of an odd-fold repetition of a code is not so obvious, however. One can easily see that

$$t((2e + 1)C) \geq en + t(C),$$

where n is the length of the linear code C. This bound is sometimes attained, as when C is itself $2C'$ for some code C' (or when $e = 0$ (!)), and more interestingly for (i) $e = 1$ and C = any cyclic code of length 7 or 9 (use [5]), and (ii) any code of dimension 2 (use [7]).

12. We discuss a final approach. Suppose we take J as a subset of an information set. Then $t(C_J) = 0$. But C_0 is a code of type $[n- \mid J \mid, k- \mid J \mid]$, so it has the same redundancy as C. Therefore the effort involved in calculating $t(C_0)$ would be roughly the same as that required to

calculate $t(C)$. Furthermore, the tendency is for $t(C_0)$ to be greater than $t(C)$, although this would be complicated to analyze. It calls to mind Remark 3. Certainly if $t(C') = t(C_0)$ then $t(C) = t(C_0)$, but the converse is not true. In fact all four possibilities (for equality or inequality) can arise in $t(C') \le t(C) \le t(C_0)$. For example, take C_0 to be the [7,3,4] simplex code, and set up $g(C)$ as

$$g(C) = \begin{array}{|c|c|} \hline 1 & v \\ \hline 0 & \\ 0 & g(C_0) \\ 0 & \\ \hline \end{array}$$

Then if $v = 0$, $C' = C_0$ and $t(C') = t(C) = t(C_0) = 3$. If $|\, v \,| = 1$, then $t(C') = 2 < t(C) = t(C_0) = 3$. If $|\, v \,| = 2$, then $2 = t(C') = t(C) < t(C_0) = 3$. If $|\, v \,| = 7$ then C' is the [7,4,3] Hamming code and C is the [8,4,4] code. Now $1 = t(C') < 2 = t(C) < 3 = t(C_0)$.

Therefore we do not attempt to bound $t(C)$ with this approach.

The Extended Direct Sum of Two Codes

We apply Theorem 1 to the code of the title, which is defined [7] by the generator matrix

$$\begin{array}{|ccc|} \hline g(L) & & 0 \\ & \ddots & \\ 0 & & g(L) \\ \hline g(B) & \cdots & g(B) \\ \hline \end{array}$$

where $g(L)$ is the generator matrix of an $[n_L, k_L]$ code L and $g(B)$ is that for an $[n_L, k_B]$ code B. If the number of copies of B here is m, then the code is of type $[n, k]$ where

$$n = mn_L, \qquad mk_L \le k \le mk_L + k_B.$$

This code may be denoted $EX(L, m, B)$, or C_m for fixed L and B.

An upper bound U on $t(C_m)$ is presented in [7, Theorem 31]. We now find by Theorem 1 another upper bound, as follows. We take for J the first pn_L coordinate-places of this code; then

$$(13) \qquad t(C_m) \;\le\; t(C_p) \;+\; (m-p)t(L)$$

for any $p = 0, 1, \ldots, m$, since the subcode of C_m 0 on J is the direct sum of $m - p$ copies of L.

This bound is too general to be useful unless we specify the codes B and L; but, perhaps surprisingly, it is helpful in the instances of this construction presented in [7].

The first of these takes L as the $[7,1,7]$ code and B as the $[7,4,3]$ Hamming code. The code C_m has type $[7m, m+3]$. There follows an excerpt from the table (79) of [7]:

m	$t(Cm)$	U
3	6	7
4	9	10
5	12 or 13	13
6	?	15
7	?	18
8	?	21
9	?	23

But our bound (13) immediately yields $t(C_5) \leq 12 = t(C_4)+3$; and it duplicates the upper bound U for $m = 6,7,8$. Ours gives, however, $t(C_9) \leq 24$ unless we can show that $t(C_m)$ is strictly less than the upper bound U for some of $m = 6,7,8$.

The second instance of the extended direct sum studied in [7] is the **Berlekamp - Gale** switching network.

This problem concerns the code C_m of the above construction when L is the $[\ell,1,\ell]$ code and B is the $[\ell,\ell,1]$ code. The code is equivalent to the space of all $m \times \ell$ matrices over Z_2 spanned by all matrices of the form

$$\begin{bmatrix} & 0 & \\ 1\ \ 1 & \ldots & 1 \\ & 0 & \end{bmatrix} \quad \text{or} \quad \begin{bmatrix} & 1 & \\ & 1 & \\ 0 & \vdots & 0 \\ & 1 & \end{bmatrix}$$

It has length ℓm and dimension $\ell+m-1$. The problem posed by Berlekamp and Gale as described in [7] comes down to this: what is the covering radius of this code?

If we take $p = m-1$ in (13) we get

(14) $$t(C_m) \leq t(C_{m-1}) + \lfloor \ell/2 \rfloor.$$

We now improve slightly the results of [7] for the code $\ell = 7$, where our bound (14) reads $t(C_m) \leq t(C_{m-1}) + 3$.

An excerpt from table (81) of [7]:

(15)

m	$t(C_m)$	U
4	8	9
5	?	12
6	?	14
7	?	16

One sees from (14) that $t(C_5) \leq 11$. We now give an ad hoc argument to reduce $t(C_5)$ further.

Proposition 16. $t(C_5) \leq 10$.

Corollary. $t(C_6) \leq 13$.

The Corollary improves the bound of (15).

Proof of the Proposition. We first study the code a bit for general ℓ. It has minimum distance $min\{\ell, m\}$. This is not hard to see; the proof is omitted as it is not a key fact for us.

The direct product of the symmetric groups $Sym(m)$ and $Sym(\ell)$ acts on this code. If m and ℓ are relatively prime, the code is equivalent to a cyclic code. Every codeword is equivalent to the codeword $v(i,j)$ for appropriate i, j, where

$$v(i,j) = j \begin{array}{|c|c|} \multicolumn{1}{c}{} & \multicolumn{1}{c}{i} \\ \hline 1 & 0 \\ \hline 0 & 1 \\ \hline \end{array} = \begin{array}{|c|c|} \hline I(j,i) & 0 \\ \hline 0 & I(m-j, l-i) \\ \hline \end{array}$$

for $0 \leq i \leq \ell$ and $0 \leq j \leq m$. By $I(x,y)$ we mean the $x \times y$ matrix of all 1's. Since $v(i,j)$ is itself equivalent to $v(\ell - i, m - j)$, a central symmetry in the matrix, the $v(i,j)$ represent all the non-0 codewords if $0 \leq i \leq \lfloor \ell/2 \rfloor$ (or $0 \leq j \leq \lfloor m/2 \rfloor$).

An information-set of $m + \ell - 1$ coordinate places may be taken as the union of the first row and first column.

A basis for the code $C_m^{\perp}$ consists of all vectors of weight 4 of the form

$$\begin{array}{c|cc|} \multicolumn{1}{c}{} & \multicolumn{1}{c}{} & \multicolumn{1}{c}{i} \\ \cline{2-3} & 1 & 1 \\ j & 1 & 1 \\ & & \\ \cline{2-3} \end{array}$$

for $2 \leq i \leq \ell$ and $2 \leq j \leq m$.

There are 24 classes of vectors $v(i,j)$ for our case; since $\mid v(i,j) \mid = ij + (\ell - i)(m - j)$ we work out the weights as

$j\backslash i$	0	1	2	3	4	5	6	7
0	35	30	25	20	15	10	5	0
1	28	25	22	19	16	13	10	7
2	21	20	19	18	17	16	15	14

No coset leader has a row of weight 4 or more nor a column of weight 3 or more.

To prove that $t(C_5) \leq 10$, we show that no vector of weight 11 is a coset leader. Such a vector b must have at least one row of weight 3; we take it as the first row (by use of the group) and denote the remaining subparts as in this diagram:

$$b = \begin{array}{|ccc|cccc|}\hline 1 & 1 & 1 & 0 & 0 & 0 & 0 \\ \hline & A & & & B & & \\ \hline \end{array}$$

Here A is a 4×3 matrix and B is a 4×4 matrix. Vertical bars denotes its weight as a vector, and an overbar denotes the complementary matrix.

Case 1. $|A| \leq 1$. Then $|B| \geq 7$. Now $b + v(3,1)$ has weight $|A| + |\overline{B}| \leq 10$.

Case 2. $|A| = 2$. Then $|B| = 6$, and

(2.1) at least one column of B is 0, or

(2.2) two columns of B have weight 1.

We compute $b + v(3,1)$ of weight 12; in the subcase (2.1) we add a column of 5 1's to produce weight 9; in the subcase (2.2) we add two such columns to reduce the weight to 10.

Case 3. $|A| = 3$. Then $|B| = 5$. The column-weights in B are $2,2,1,0$ or $2,1,1,1$. Adding $v(3,1)$ we get weight 14, and the column-weights in $\overline{B}$ are 2,2,3,4 or 2,3,3,3. Adding two columns of weight 5 reduces the first subcase to weight 10. The second subcase is more difficult. We treat it in a series of further cases.

Observe that when $|A| = 3$, by using the group we may take A to be one of the following three 4×3 matrices:

(16)
$$\begin{array}{|ccc|}\hline 1 & 1 & 1 \\ & 0 & \\ \hline \end{array} \qquad \begin{array}{|ccc|}\hline 1 & 1 & 0 \\ 0 & 0 & 1 \\ & 0 & \\ \hline \end{array} \qquad \begin{array}{|ccc|}\hline 1 & 0 & \\ 0 & 1 & 0 \\ 0 & 0 & 1 \\ & 0 & \\ \hline \end{array}$$

Case 4 Case 5

In the first of these we add $v(3,2)$ to b to get a vector of weight 7. In the second, Case 4, and third, Case 5, we have more work to do.

Case 4. A is the second matrix in (16), and B has column-weights 2,1,1,1. We add $v(3,2)$ to b, getting weight 9 if B_1 is 0 and weight 11 if $|B_1| = 1$, the only two possibilities, where B_1 is the top row of B.

If $|B_1| = 1$ we use the group to arrange b as

$$b = \begin{array}{|ccc|cccc|}\hline 1 & 1 & 1 & 0 & 0 & 0 & 0 \\ 1 & 1 & 0 & 1 & 0 & 0 & 0 \\ 0 & 0 & 1 & p & q & r & s \\ \hline 0 & 0 & 0 & \multicolumn{1}{c|}{w_1} & & w_2 & \\ 0 & 0 & 0 & \multicolumn{1}{c|}{} & & & \\ \hline \end{array} \qquad \begin{array}{l} w_1 = \text{wt of } 2 \times 1 \text{ block} \\ w_2 = \text{wt of } 2 \times 3 \text{ block} \end{array}$$

We add $v(4,3)$ to b to get a vector of weight

$$4+\overline{p}+6-w_2+w_1+q+r+s=10+q+r+s-w_2+w_1+\overline{p}. \tag{17}$$

If $p=1$, then at most one of q,r,s is 1 and $w_1=0$ and $w_2\geq 2$. This gives weight $\leq 11-2=9$. If $p=0$ and two of q,r,s are 1, then we renormalize b on row 3. That is, we put the three 1's of row 3 in the first row and first three columns. This yields the case $|A|\leq 2$. If one of q,r,s is 1, then (17) yields 9 if $w_1=0$ and 11 if $w_1=1$. In the latter case we have

$$b=\begin{array}{|ccccccc|}\hline 1&1&1&&&&\\ 1&1&0&1&0&0&0\\ 0&0&1&0&1&0&0\\ 0&0&0&1&0&1&0\\ 0&0&0&0&0&0&1\\ \hline\end{array}$$

and $b+v(3,3)$ has weight 10. This settles Case 4.

<u>Case 5</u>. A is the third matrix in (16), and B has column-weights $2,1,1,1$. Now b is

$$b=\begin{array}{|ccccccc|}\hline 1&1&1&0&0&0&0\\ 1&0&0&&&&\\ 0&1&0&&&&\\ 0&0&1&&&B&\\ 0&0&0&&&&\\ \hline\end{array}$$

If B has a row of weight 3, we renormalize b on that row to get an A of weight 0 or 1. If B has a row of weight 2 not the last row, we again renormalize b and reduce to a previously settled case. The only remaining possibility is that the last row of B has weight 2, and that the other rows of B each have weight 1.

Using the group again we may put b in the form

$$b=\begin{array}{|ccccccc|}\hline 1&1&1&0&0&0&0\\ 1&0&0&1&0&0&0\\ 0&1&0&0&1&0&0\\ 0&0&1&0&0&1&0\\ 0&0&0&0&0&1&1\\ \hline\end{array}$$

Now $b+v(5,3)$ has weight 10.

This completes the proof that $t(C_5)\leq 10$.

The $u \mid u+v$ Construction

The $u \mid u+v$ construction applies to nonlinear codes C_1 and C_2. We can generalize the bound of Theorem 1 to the nonlinear situation as follows: Let C_1 be an (n,M) code, K an

(m, M) code, and C_2 an (m, M') code. Let β be a bijection, $\beta: C_1 \longrightarrow K$. Define a code C as $\{(u, \beta(u)+v); u \in C_1, v \in C_2\}$.

THEOREM 18. C is an $(n+m, MM')$ code such that

$$t(C) \leq t(C_1) + t(C_2).$$

In the $u \mid u+v$ construction, $K = C_1$ and β is the identity. The proof of this theorem, omitted here, is the same as in the linear case.

Schatz [17; 3, §IV] found a special case of this bound for Reed - Muller codes, which have a recursive genesis via the $u \mid u+v$ construction [12, p. 374].

We can apply Theorem 18 to the nonlinear codes of [6, 10, 18]. We begin by remarking that each of these codes C_8, C_9, C_{10}, C_{11} has covering radius 2. These codes have parameters

$$\begin{aligned} C_8&: \quad (8,20,3) \\ C_9&: \quad (9,38,3) \\ C_{10}&: \quad (10,72,3) \\ C_{11}&: \quad (11,144,3) \end{aligned}$$

These codes are all derived from a Steiner system $S(5,6,12)$, a set of 132 6-subsets, called 6-*clubs*, of a 12-set X, with the property that every 5-subset of X is contained in exactly one 6-club. The reader may find the definitions in [18].

It is a routine exercise to verify that for each of these codes and for each w, every vector of weight w in the ambient space is within distance 2 of some codeword. We omit the proof except for these remarks:

- Every 6-subset of X not a 6-club meets 6 different 6-clubs in 5 points. That is, if $A = \{a_1, \ldots, a_6\} \subseteq X$, and if $\{b_1, \ldots, b_6\} = X - A$, then the notation can be chosen so that $(A - \{a_i\}) \cup \{b_i\}$ is a 6-club for each i. This simple observation is helpful in the above verification.
- Every 7-subset of X contains exactly one 6-club. Every 8-subset contains exactly four 6-clubs.
- We show that every vector of weight 5 is within distance 2 of C_8. (This is the least obvious of the proofs.) This code consists of 0^8, 1^8, $(01)^4$, $(10)^4$, and all vectors of weights 3 and 4 projected onto the first 8 coordinate-places from 6-clubs as in the diagram

$$\begin{array}{ccc} 1 \quad \ldots \quad 8 & & 9 \quad 10 \quad 11 \quad 12 \\ \hline \ldots \; 1 \; 1 \; 1 \; \ldots & & 1 \quad 1 \quad 1 \quad 0 \\ \ldots 1 \; 1 \; 1 \; 1 \ldots & & 0 \quad 1 \quad 1 \quad 0 \end{array},$$

where the column-headings indicate the points of X, and the rows are 6-clubs. That is, only the 6-clubs containing 10 and 11 and not 12 give rise to words of C_8. We will show that every 5-set F (on the first 8 points) contains a codeword of C_8.

Consider the 7-sets consisting of $\{9,10,11\} \cup F_i$, where F_i is F less the ith point of F, for $i = 1, \ldots, 5$. Each of these contains a unique 6-club G_i.

Now consider the 8-set $E = F \cup \{9,10,11\}$ and its 4 6-clubs. To each 6-club H is associated a 2-subset of E, namely $E - H$. These 4 2-subsets are mutually disjoint. Therefore one of them is disjoint from $\{9,\ 10,\ 11\}$, so F contains the associated word of weight 3 from C_8. And this happens to be true even if F contains a codeword of weight 4.

• We remark also that the 8 vectors of weight 3 in C_8 cover $8 \times 3 = 24$ 2-subsets of the 8 coordinate-places; but $\binom{8}{2} = 28$, so there are vectors of weight 2 at distance 2 from C_8.

The $u \mid u+v$ construction is applied in [18] to $C_1 = E_i$, the $[i, i-1, 2]$ code of all even-weight vectors of length i, and $C_2 = C_i$ $(i = 8, \ldots, 11)$.

We now find the covering radius for all such constructions:

Theorem 19. Let C be any (n, M) code and E the $[n, n-1, 2]$ code. Then $t(C \mid C + E) = 1 + t(C)$, and that of $t(E \mid E + C) = t(C)$ if $t(C) \geq 2$. If $t(C) = 1 = d(C)$ and every codeword is at distance 1 from some codeword, then $t(E \mid E + C)$ is also $1 = t(C)$. Otherwise $t(E \mid E + C) = 1 + t(C)$.

Proof. That $1 + t(C)$ is an upper bound follows from Theorem 18.

For $C \mid C + E$: Let v be any n-vector at distance $t(C)$ from C. All the vectors in C at distance $t(C)$ from v have weights of the same parity p. Let u be any n-vector with weight of parity opposite to p. Then (v, u) is at distance $1 + t(C)$ from $C \mid C + E$.

For $E \mid E + C$: $t(C)$ is a lower bound on the covering radius because the distance of $(0, v)$ to the code is the minimum of the weights $w = \mid (e, e + c + v) \mid$ where $e \in E$, $c \in C$. If v is at distance $t(C)$ from C, then

$$w = \mid e \mid + \mid e + (c + v) \mid \ \geq \ \mid e \mid + t(C) - \mid e \mid \ = \ t(C).$$

Now $t(C)$ (if ≥ 2) is an upper bound because for any n-vectors x, y, the vector (x, y) is within distance $t(C)$ of $E \mid E + C$, as we now show.

$$(x, y) + (x + u, x + u + c) = (u, x + y + u + c) = z$$

where we take $u = 0$ if $\mid x \mid$ is even, and u as any unit vector if $\mid x \mid$ is odd; c is any word of C. When $u = 0$, z has weight at most $t(C)$ for some c. When $u \neq 0$, z has weight $1 + \mid (x+y)+c+u \mid$. If $x + y \notin C$, then minimize $\mid x + y + c \mid \leq t(C)$ and choose $u \in x + y + c$. Then $\mid z \mid \leq t(C)$. If $x + y \in C$, then $\min \mid z \mid \leq 2$.

Still in the case $x + y \in C$, take now $t(C) = 1$; if for all $c' \in C$ there is a $c \in C$ such that $\mid c' + c \mid = 1$, then choose such c for $c' = x + y$ and set $u = c + c'$. This makes $\mid z \mid = 1 = t(C)$. If $d(C) = 1$ but C does not have this "uniformity" property, or if $d(C) > 1$ then $t(E \mid E + C) = 2 = 1 + t(C)$.

Finally, if $t(C) = 0$, then $C = Z_2^n$; and $E \mid E + C$ is the Cartesian product $E \times Z_2^n$ with covering radius $1 = 1 + t(C)$. QED.

The codes $C_8, \ldots, C_{11}$ are used (in [18, Thm. 2]) to construct an infinite family of nonlinear codes C_n of length n and distance 3 for all n satisfying $2^m \leq n < 3.2^{m-1}$, and all $m \geq 3$.

The construction is defined inductively. C_n for $n = 8, \ldots, 11$ are defined above. If $m \geq 4$ and $n = 2j$ is even, then $C_n = E_j \mid E_j + C_j$. If $n = 2j+1$ is odd, then $C_n = (E_{j+1} \mid E_{j+1} + C_j^*)'$, where C^* denotes the code obtained from C by the appending of an overall parity check; and C' denotes the puncturing of C on one coordinate.

Proposition 20. Let m and n be integers such that $2^m \leq n < 3.2^{m-1}$ and $m \geq 3$. With C_n defined as above,

$$t(C_n) = 2.$$

Proof. For even n the result follows immediately from Theorem 19. For odd n we use the same theorem and the fact [1,3] that appending an overall parity-check always increases the covering radius by 1, and that puncturing an even code on one coordinate always reduces it by 1.

Results on Specific Codes

We now apply the foregoing ideas to specific codes. Some of these covering radii have been determined by exhaustion [5], and we have been able to get the same result more simply. As we shall see, the present approach fails to produce the exact value in some cases.

The results are presented in Table 1. They are explained afterwards. In most of these codes the bound of Theorem 1 is better than the Delsarte bound [4].

TABLE 1

n	k	d	$t(C)$	Upper bound		
15	8	3	3	3	Cor 5	Random
15	8	2	3	3	Cor 7	Random
15	8	3	3	3	Cor 5	Random
15	8	2	4	4	Cor 7	Random
15	8	4	4	4	Cor 5	Nonrandom
17	8	6	5	5	Cor 4	Cyclic
17	9	5	3	4	Cor 5	Cyclic
21	8	6	7	7	Cor 7	Cyclic; roots $0, 1, 3, 9$
31	15	6	9	9	Cor 4	Cyclic; roots $0, 1, 3, 7$
31	15	8	6	7	Cor 5	Cyclic; roots $0, 1, 5, 11$
31	15	8	7	8	Cor 5	QR; roots $0, 1, 5, 7$
31	15	8	7	8	Cor 5	BCH; roots $0, 1, 3, 7$
41	20	10	?	11	Thm 10	lower bound 9; QR
47	23	12	?	12	See text	lower bound 11; QR

We first exhibit the results of Corollaries 5 and 7 applied to four mutually inequivalent [15,8] codes chosen at random in the form $g(C) = I_8, D$, where the 8 rows of D were chosen by the APL function "deal" as distinct integers from 1 to 127. Thus $d(C) \geq 2$. In each case the first choice of $\overline{J}$, as a codeword of weight 3, yielded $t(C_J) = t(C) - 1$. In the first three cases $t(C_J) = 2$, giving $t(C) \leq 3$. No further calculation is necessary to determine $t(C)$, because $t[15,8] = 3$ [3, 7]. In the fourth (and fifth) cases we got $t(C) \leq 4$, so we ran a covering radius program on C to the point where it reported that $t(C) \geq 4$.

The fifth case is a nonrandomly chosen [15,8,14] code. D is the 7 cyclic shifts of $1^3 0^4$, with 11010^3 as a last row. Choosing J as the complement of row 1 of $g(C)$ immediately yielded $t(C_J) = 2$, hence $t(C) \leq 4$.

The [17,8,6] cyclic code C has weights 0,6,8,10,12. Therefore Corollary 4 applied to a vector of weight 10 ($s = 1$) immediately yields $t(C) \leq 5$. Since the [17,9,5] cyclic code contains $C, 5 \leq t(C)$ by the supercode lemma [3].

These methods, however, cannot yield a better upper bound than 4 for the [17,9,5] cyclic code. The reason lies in the tables of $t[n,k]$ and $d[n,k]$ and is similar to that for the [31,16,7] code, explained below.

The [21,8,6] cyclic code C with roots 0,1,3,9 has a codeword v of weight 10. Taking $J = \overline{v}$ we get $t(C_0) = 5$, and it then becomes a simple matter to calculate $t(C_J) = 2$, since C_J has redundancy 4 as an [11,7] code. Therefore $t(C) \leq 7$, which is again a lower bound by the supercode lemma. That is, $C \oplus 1^{21}$ has minimum odd weight 7 ([5]) since the maximum weight in C is 14 [2].

We use Corollary 4 on the [31,15,6] cyclic code C with roots 0,1,3,7. We knew that $9 \leq t(C) \leq 12$[14], but now we can show that $t(C) \leq 9$: We find a codeword J of weight 6 and then the weight-distribution $\{n_i\}$ of the subcode C_0 of all codewords 0 on J. It is, in part, $n_6 = 10$, $n_{12} = 140$. Therefore C_0 has a vector of weight 12 which is not the sum of two codewords of weight 6. By Corollary 4, $t(C) \leq 9 = \lceil 5/2 \rceil + \lceil 11/2 \rceil$.

The [31,15,18] cyclic code C with roots 0,15,11 is next. An easily found codeword $\overline{J}$ of weight 8 yielded $t(C_J) = 3$ for the [23,14] code C_J. Thus $t(C) \leq 8/2+3 = 7$. Since the supercode $C \oplus 1^{31}$ has $d = 5$ we can say $5 \leq t(C) \leq 7$. The true value is 6 [5].

It appears to be impossible to improve the upper bound 7 because of considerations of $t[n,k]$; cf. "Limitations" below.

For both the [31,15,8] quadratic-residue and BCH codes the generator polynomials have minimum weight. With $\overline{J}$ as this vector, C_J is a [23,9] code, and $t(C_J) = 4$ in both cases. Hence our upper bound is 8, from Cor. 5. All minimum-weight vectors are equivalent under the group of the QR code; in the BCH, the group generated by the cyclic shift and the squaring operation has order $31 \cdot 5$. In each of the three orbits of minimum-weight vectors, $t(C_J)$ was at least 4.

The [41,20,10] code C has $9 \leq t(C) \leq 11$ from the supercode lemma and from Cor. 7 or Theorem 10. A codeword $\overline{J}$ of weight 20, found at random, has no descendants. C_J is a [21,19] code; we find $t(C_J) = 1$ by inspection. The calculations here involve only the finding of the vector of weight 20 and row operations on $g(C_J)$.

The [47,23,12] code C has $11 \leq t(C) \leq 12$, the "11" coming from the supercode lemma. Here we get the upper bound 12 by a searching method for finding J. Taking $g(C)$ in the form I_{23}, D, we select various subsets of three rows, say rows numbered a, b, c, such that the 3×24 submatrix $D*$ of D on those three rows has a "large" number of columns either (000)′ or (111)′. In practice we found a, b, c making this number 9. We then take J to be all the columns (000)′ or (111)′ of D^* and of I_{23} except for the columns a, b, c because the 2-dimensional subcode spanned by $R_a + R_c$ and $R_b + R_c$ is precisely C_0 for that choice of J. (R_i denotes row i). The situation is depicted in Figure 4; the columns have been permuted for clarity.

C_J is now a [29,21] code for which the calculation of $t(C_J) = 3$ by exhaustion is quite feasible in APL, and C_0 is an [18,2] "doubled" code (see 11) of covering radius 9. In general there is a simple formula in [7] for the covering radius of any 2-dimensional code.

The Delsarte [4] upper bound on $t(C)$ is 15.

	a	b	c	I_{23} less a,b,c			D*				
R_a	1	0	0	0	. . .	0	0	0	1	1	
R_b	0	1	0	0	. . .	0	0 . . .	0	1 . . .	1	*
R_c	0	0	1	0		0	0	0	1	1	
		3			20				9		15
						J					
R_a+R_c	1	0	1	0	. . .	0	0 . . .	0	0 . . .	0	*
R_b+R_c	0	1	1	0		0	0	0	0	0	

Figure 4.

Limitations of Theorem 1

Theorem 1 does not always produce the true covering radius. For example, the [31,16,6] cyclic code C, with roots 1,3,7, has covering radius 5 [5], but one can see that any application of Theorem 1 to C would yield for $t(C)$ an upper bound larger than 5. To prove this claim we focus on C_0, an $[n, k, d]$ code with $d \leq 6$, and on $C_J^{\perp}$, (see) a code of type $[31 - n, 15 - n + k, d']$ in which d' is necessarily at least 8 because $C_J^{\perp}$ is a shortened subcode of $C^{\perp}$, a [31,15,8] code (see [5] or [16]).

CASE 1. $C_J^\perp$ has positive dimension. "Because of C_0" $n \leq 6$; "because of $C_J^\perp$," $n \leq 23$. We therefore find the largest value of k (from [9]) for which there is both an $[n, k, \leq 6]$ code and a $[31 - n, 15 - n + k, \leq 8]$ code for $n = 6, \ldots, 23$. In compact form the results are

$n =$	6	9	11	12	15	19	21	22	23
$k \leq$	1	2	3	4	5	6	7	8	9

This means, e.g., if $n = 8, k \leq 1$; if $n = 18, k \leq 5$, etc. Now we find the associated values $31 - n, 16 - k$ for C_J. Finally we look up $t[n, k]$ and $t[31 - n, 16 - k]$ for each of these values of n, k, using [3] or [7]. In each case we find that $t(C_0) + t(C_J) \geq 6$.

CASE 2. $C_J^\perp$ has dimension 0. In this case C_J is the full code $Z_2^{|J|}$ with covering radius 0; i.e., J is contained in an information set. Then we would be looking at codes C_0 of redundancy $n - k$: $[31 - j, 16 - j]$ for $j - 1, \ldots, 16$.

$$t[31 - j, 16 - j] = \begin{cases} 5 & \text{for } j = 1, \ldots, 4 \\ 5 \text{ or } 6 & j = 5, \ldots, 9 \\ 7 \text{ or more} & j = 10, \ldots, 16. \end{cases}$$

The computation of $t(C_0)$, however, since C_0 has the same redundancy as that of C, would be only marginally less burdensome for the lengths near 31 that we would need to try. And of course we might have to try many cases before finding one that yielded covering radius 5.

But Theorem 1 easily determined the covering radius of the even-weight subcode of C.

Another code, the [31,16,5] cyclic code B (having roots 1,15,11), has $d(B^\perp) = 6$. Another tabular analysis like the above suggests only one hopeful possibility: to choose B_0 as a [5,1,5] code attached to a minimum-weight vector. Since there are only 31 vectors of weight 5, it suffices to try any one of them. The result: $t(C_0) = 2$, but $t(B_J) = 5$. Since $t(B) = 5$ [5], we fail in this case, too. (Some other remote possibilities are also suggested by the table; they involve parameters for B_0 or B_J for which the minimum covering radius is unknown. For example, if we could choose B_0 as a [14,6,5] code with $t(B_0) = 3 = t[14, 6]$, then B_J would be a [17,10] code. And if $t(B_J) = 2$ — thus providing a new entry for the tables of $t[n, k]$ in [3,7], since at present $t[17, 10]$ is known only to be 2 or 3 — we would have the result. But it's very "iffy." Even the existence of such a B_0 is doubtful, since it could have no vectors of weights 6 or 7 [16].

Conclusion

Theorem 1 and its corollaries provide a simpler means of finding the covering radius for some codes than exhaustive calculations. For some other codes it provides reasonable upper bounds. There are codes, however, for which the upper bound of Theorem 1 is necessarily larger than the covering radius.

Acknowledgements

I am grateful to Neil Sloane for his question about the presentation of the first version of Theorem 1 in [3,§4.5]; it led me to the present version. I also thank H. Janwa for his observation that C_0 may be replaced by any subcode in Theorem 1. Finally, I thank my mother, Mrs. Mattson, for "acarpous."

References

1. E. F. Assmus, Jr., and Vera Pless, "On the covering radius of extremal self-dual codes," IEEE Trans. Inform. Theory, IT-29 (1983) 359–363.
2. E. R. Berlekamp *Algebraic Coding Theory*, McGraw Hill, New York, 1968.
3. Gérard D. Cohen, Mark R. Karpovsky, H. F. Mattson, Jr., and James R. Schatz, "Covering Radius—Survey and Recent Results," IEEE Trans. Inform. Theory IT-31 (1985), 328–343.
4. Ph. Delsarte, "Four fundamental parameters of a code and their combinatorial significance," Inform. and Control 23 (1973), 407–438.
5. Diane E. Downey and N. J. A. Sloane, "The covering radius of cyclic codes of length up to 31," IEEE Trans. Inform. Theory, IT-31 (1985), 446–447.
6. M. J. E. Golay, "Binary coding," IEEE Trans. Inform. Theory, PGIT-4 (1954) 23–28.
7. R. L. Graham and N. J. A. Sloane, "On the covering radius of codes," IEEE Trans. Inform. Theory, IT-31 (1985), 385–401.
8. J. H. Griesmer, "A bound for error-correcting codes," IBM J. Res. Develop. 4 (1960), 532–542.
9. H. J. Helgert and R. D. Stinaff, "Minimum-distance bounds for binary linear codes, " IEEE Trans. Inform. Theory IT-19 (1973), 344–356.
10. D. Julin, "Two improved block codes," IEEE Trans. Inform. Theory, IT-11 (1965) 459.
11. Mark R. Karpovsky, public communication, at Journée sur le rayon de recouvrement et codes correcteurs d'érreurs, ENST, Paris, 26 June 1984.
12. F. J. MacWilliams and N. J. A. Sloane, *The Theory of Error-Correcting Codes*, North Holland, Amsterdam, 1977.
13. H. F. Mattson, Jr., "An upper bound on covering radius," Annals of Discrete Math. 17 (1982) 453–458.
14. H. F. Mattson, Jr. "Another upper bound on covering radius," IEEE Trans. Inform. Theory, IT-29 (1983) 356–359.
15. W. W. Peterson and E. J. Weldon, Jr., Error-correcting codes, Second Edition, Cambridge, M.I.T. 1972.
16. V. Pless and E. A. Prange, "Weight distribution of all cyclic codes ... [of length] 31 over GF(2)" unpublished memorandum, September, 1962.
17. James R. Schatz, "On the coset leaders of Reed-Muller codes," Ph. D. dissertation, Syracuse University, 1979.
18. Neil J. A. Sloane and Donald S. Whitehead, "New family of single-error correcting codes, IEEE Trans. Inform. Theory, IT-16 (1970) 717–719.

ASSOCIATION SCHEMES AND DIFFERENCE SETS DEFINED ON TWO WEIGHT CODES

Llorenç Huguet and Mercè Griera
Departament d'Informàtica, Facultat de Ciències,
Universitat Autònoma de Barcelona, Bellaterra,
Barcelona, España.

In this paper we establish the equivalence between the association scheme defined over the cosets of the orthogonal of a two weight linear code and the difference set defined on Ω^*, the column set of the generator matrix of the same two weight code.

I.- INTRODUCTION

Association schemes have already been considered by Delsarte (1) under algebraic theory point of view. In (1) a deep connection between s-weight linear codes C, coset-weight enumerator of $C^\perp$ and s-classes association schemes is established. In particular, all coset-weight enumerators of a completely regular code $C^\perp$ are characterized in terms of the eigenmatrix rows associated to s-weight linear code C.

Partial difference sets have already been considered by Camion (2) and Wolfmann (3). In (3) (see also (4)) a deep connection between two-weight codes, uniformly packed codes and difference sets is established.

In this paper we establish the equivalence between the association scheme associated to a 2-weight linear code and the partial difference set associated to column set Ω^* of the same code. In particular, we give the parameters of the partial difference set in terms of association scheme's parameters.

II.- DEFINITIONS AND USEFUL RESULTS

Let $F_q = GF(q)$ be the q elements Galois Field, being q a power of a prime p. Set $F_q^* = F_q - \{0\}$.

II.1.- DEFINITION

A symmetric association scheme with n-classes (or relations) consists of a finite set X of points together with n+1 relations $R_0, R_1, \ldots R_n$ defined on XxX which satisfy:

i) Each R_i is symmetric.
ii) For every $(x,y) \in X\text{x}X$; $(x,y) \in R_i$ for exactly one i.
iii) $R_0 = \{(x,x) : x \in X\}$ is the identity relation.

iv) If $(x,y)\varepsilon R_k$, the number of $z\varepsilon X$ such that $(x,z)\varepsilon R_i$ and $(y,z)\varepsilon R_j$ is a constant p_{ij}^k depending on i,j and k but not on the particular choice of x and y.

II.2.- REMARK

Let D_i be the adjacency matrix of the graph (X,R_i), for i=0,1,...,n. Then the commuting symmetric matrices $D_0,D_1,\ldots,D_n$ span a (n+1)-dimensional real algebra $(D_iD_j=D_jD_i=\sum_{k=0}^{n} p_{ij}^k D_k)$; called the Bose-Mesner Algebra of the association scheme.

II.3.- DEFINITION

The Hamming scheme H(n,q) is an association scheme with n classes where $X=F_q^n$ and a pair of vectors (x,y) is in R_i if and only if the Hamming distance d(x,y)=i.

II.4.- DEFINITION

Let $(X,R=\{R_i\})$ and $(X',R'=\{R_i'\})$ be two association schemes. We define an isomorphism between these two association schemes as an isomorphism $\sigma: X \longrightarrow X'$ such that for $R_i\varepsilon R$ then $\sigma R_i=\{(\sigma x,\sigma y)/\ (x,y)\varepsilon R_i\}\varepsilon R'$.

II.5.- DEFINITION

A subset $D\subset F_q^k$ is said to be a partial difference set with two parameters μ_1 and μ_2 if

1) $F_q^* D=D$

2) $\mathrm{Card}\ \{(a_1,a_2)\varepsilon DxD\colon h=a_1-a_2\}=\begin{cases}\mu_1 & \text{if } h\varepsilon D-\{0\}\\ \mu_2 & \text{if } h\varepsilon D^C-\{0\}\end{cases}$

where $D^C=F_q^k-D$ is the complement of D.

The proof of our main theorem assumes results established by Delsarte (rf. (1)).

II.6.- THEOREM

Let C a s-weight linear code whose orthogonal code has minimum distance d satisfying $2s-1\leq d\leq 2s+1$. Then the restriction of the Hamming scheme on C is an association scheme with s-classes, say (C,R^C). Moreover: $R_i^C=\{(x,y)\varepsilon CxC\colon d(x,y)=w_i\}$, i=0,1,...,s being $w_0=0$ and $w_1,\ldots,w_s$ the s non-zero weights of C.

Set $Y=F_q^n/C^\perp$. Then we can consider that C is isomorphic to the coset decomposition of $C^\perp$; say $C\simeq Y=\{Y_0,Y_1,\ldots,Y_N\}$ $(N=q^k-1)$, where $Y_i=C^\perp+u_i$.

With the same assumption of theorem II.6 we can define a partition on YxY, say $R^Y=\{R_0^Y,R_1^Y,\ldots,R_S^Y\}$, such that $R_i^Y=\{(Y_i,Y_j)\varepsilon YxY\colon d_{min}(Y_i,Y_j)=i\}$ for i=0,1,...,s being $d_{min}(Y_i,Y_j)$ the minimum Hamming weight in the coset $C^\perp+(u_i-u_j)$.

II.7.- THEOREM (see also rf. (5))

1) (Y,R^{Y}) is an association scheme with s-classes.

2) (Y,R^{Y}) is isomorphic to the dual association scheme of (C,R^{C}).

III.- DIFFERENCE SETS AND ASSOCIATION SCHEMES DEFINED ON A s-WEIGHT CODES

Let C be a linear projective code with length n and dimension k, defined over F_q. Let G_{kxn} be the generator matrix of C and consider the column set Ω^* of G_{kxn}. Take $D^*=F_q^*.\Omega^*$

If C has s non-zero weights then we can define the following partition $R^{D^*}=\{R_0^{D^*},R_1^{D^*},\dots,R_s^{D^*}\}$ on $F_q^k xF_q^k$, such that:

$$R_0^{D^*}=\{(x,x)\varepsilon F_q^k xF_q^k\}$$

and for $i\leq s$ $\quad R_i^{D^*}=\{(x,y)\varepsilon F_q^k xF_q^k\text{: } y-x=g_{j_1}+\dots+g_{j_i}\text{ ; } g_{j_k}\varepsilon D^*$ and

$$y-x\neq g_{t_1}+\dots+g_{t_1}\text{ if } 1<i\}$$

III.1.- THEOREM

(F_q^k,R^{D^*}) is an association scheme with s-classes isomorphic to association scheme (Y,R^{Y}).

Proof:

We can define the isomorphism $s:Y\longrightarrow F_q^k$

$$C^{\perp}+u_i\longrightarrow s(u_i)=G.u_i$$

where $s(u_i)$ is the syndrome of the coset leader u_i.

If the minimum weight of $C^{\perp}+(u_i-u_j)$ is k then there exists a vector $u\varepsilon C^{\perp}+(u_i-u_j)$ such that:

$$u=a_1e_{j_1}+a_2e_{j_2}+\dots+a_ke_{j_k}$$

where $a_i\varepsilon F_q^*$ and $e_{j_i}\varepsilon\{e_1,e_2,\dots,e_n\}$ the canonic base of F_q^n. Since $s(u_i-u_j)=s(u)=$ $=a_1g_{j_1}+\dots+a_kg_{j_k}$ then it follows that $s(R_k^{Y})=R_k^{D^*}$ and the theorem is proved.

III.2.- THEOREM

If C is a two-weight projective linear code then $D=F_q.\Omega^*(=\{0\}\cup D^*)$ is a difference set with parameters μ_1 and μ_2 if and only if (F_q^k,R^{D^*}) is an association scheme with two classes.

Moreover: $\mu_1=p_{11}^1+2$ and $\mu_2=p_{11}^2$

Proof:

Suppose that (F_q^k,R^{D^*}) is an association scheme with two classes. Considering $F_q^k=\{0\}\cup D^*\cup(F_q^k-D)$ we can write the first row of the adjaceny matrix D_0,D_1 and D_2 of the scheme as:

$$D_0(0,y)=\begin{cases}1 & \text{if } y=0\\ 0 & \text{otherwise}\end{cases},\quad D_1(0,y)=\begin{cases}1 & \text{if } y\varepsilon D^*\\ 0 & \text{otherwise}\end{cases},\quad D_2(0,y)=\begin{cases}1 & \text{if } y\varepsilon F_q^k-D\\ 0 & \text{otherwise}\end{cases}$$

Because $D_1.D_1=p_{11}^0D_0+p_{11}^1D_1+p_{11}^2D_2$ we can write the first row of $D_1.D_1$ as:

$(D_1.D_1)(0,y)=p_{11}^0$ if $y=0$, $(D_1.D_1)(0,y)=p_{11}^1$ if $y\varepsilon D^*$, $(D_1.D_1)(0,y)=p_{11}^2$ if $y\varepsilon F_q^k-D$

Computing this product of matrices we obtain:

$$p_{11}^0 = \text{card } D^* \qquad p_{11}^1 = \text{card } \{z\varepsilon D^*:\ z-y\varepsilon D^*\} \quad \text{for all } y\varepsilon D^*$$

$$p_{11}^2 = \text{card } \{z\varepsilon D^*:\ z-y\varepsilon D^*\} \quad \text{for all } y\varepsilon F_q^k-D$$

If we call $t_h=$ card $\{(a_1,a_2)$ DxD: $h=a_1-a_2\}$ then we have:

$t_h=$ card $\{(a_1,a_2)\varepsilon D^*xD^*: h=a_1-a_2\}$ + $\{(a,0)$ or $(0,a):h=a\}$ and consequently $t_h=p_{11}^1+2$ if $h\varepsilon D^*$ or $t_h=p_{11}^2$ if $h\varepsilon F_q^k-D$. That is, D is a difference set with parameters $\mu_1=p_{11}^1+2$ and $\mu_2=p_{11}^1$.

The converse has been pointed out in (2) and (6), that is, if D is a difference set and we take Ω^* being a subset of D obtained by choosing a nonzero vector in each one dimensional subspace contained in D then the subspace $C(\Omega^*)$ generated by the rows of a matrix whose column set is Ω^* is a projective code with two non-zero weights.

Since the minimum Hamming distance of C' is at least 3, theorems II.6, II.7, and III.1 prove that (F_q^k,R^{D^*}) is an association scheme with two classes.

IV.- EXAMPLE

$\Omega^*=\{0010,0110,1011,0100,0101,1110\}$ is the column set of a projective linear code with nonzero weights $w_1=2$ and $w_2=4$. (in this case n=6 and k=4)

		0000	0010	0110	1011	0100	0101	1110	1000	1100	1110	0001	1001	1101	0011	0111	1111
	1111	2	2	2	1	1	2	2	2	2	1	1	1	1	2	2	o
	0111	2	1	2	2	2	1	2	2	1	2	1	1	2	1	o	2
	0011	2	2	1	2	2	1	2	1	2	2	1	2	1	o	1	2
	1101	2	2	1	1	2	2	2	1	2	2	2	1	o	1	2	1
F_2^4-D	1001	2	1	2	1	2	2	2	2	1	2	2	o	1	2	1	1
	0001	2	2	2	2	1	1	2	2	2	1	o	2	2	1	1	1
	1010	2	2	2	2	1	2	1	1	1	o	1	2	2	2	2	1
	1100	2	1	2	2	2	2	1	1	o	1	2	1	2	2	1	2
	1000	2	2	1	2	2	2	1	o	1	1	2	2	1	1	2	2
	1110	1	2	2	1	2	1	o	1	1	1	2	2	2	2	2	2
	0101	1	2	2	1	2	o	1	2	2	2	1	2	2	1	1	2
	0100	1	1	1	2	o	2	2	2	2	1	1	2	2	2	2	1
$\Omega^*=D^*$	1011	1	2	2	o	2	1	1	2	2	2	2	1	1	2	2	1
	0110	1	1	o	2	1	2	2	1	2	2	2	2	1	1	2	2
	0010	1	o	1	2	1	2	2	2	1	2	2	1	2	2	1	2
	0000	o	1	1	1	1	1	1	2	2	2	2	2	2	2	2	2

This is the association scheme with two classes associated to $C(\Omega^*)$, where o,1 and 2 indicate that (x,y) is in R_o, R_1 or R_2 respectively.

Choosing $D=\Omega^* \cup \{0\}$ we have a difference set with parameters $\mu_1=4$ and $\mu_2=2$ (see (3) or (6)).

The reader will observe easily that $p^1_{11}=p^2_{11}=2$.

REFERENCES

(1) DELSARTE P. "An Algebraic Approach to the Association Schemes of Coding Theory". Philips Research Rep. Supp. 10 (1973).

(2) CAMION P. "Difference Sets in Elementary Abelian Groups". Les Presses de l'Université de Montréal, Montréal (1979).

(3) WOLFMANN J. "Codes projectifs a deux poids, "caps" complets et ensemble de differences". J. Combinatorial Theory (A) 23 (1977).

(4) HUGUET L. "Dual Weight Enumerators of the Cosets of a Regular Codes, over GF(q), and their Combinatorial Significance". I.E.E.E. Int. Symposium on Information Theory (1982).

(5) GRIERA M. "Esquemes d'associació: aplicació a la teoría de codis". These: Universitat Autònoma de Barcelona (1984).

(6) HUGUET L. "Regular Codes: Application to wire-Tap Channel". These: Universitat Autònoma de Barcelona (1981).

AUTOMORPHISMS OF TWO FAMILIES OF EXTENDED NON BINARY CYCLIC GOPPA CODES

J.A. THIONG-LY
AAECC
Lab. LSI, University P. Sabatier
118 route de Narbonne
31062 Toulouse Cedex - France.

Summary : We give automorphisms subgroups of two classes of extended non binary Goppa Codes.

INTRODUCTION

Goppa codes defined by V.D. Goppa in 1970 [2] are nowadays among the most important linear codes discovered these later years. They are a subject of many studies. However, there exists few results about the Automorphisms Group of these codes. We know :

- a result of E.R. Berlekamp and O. Moreno [1], [5], and
- a result of K.K. Tzeng and K. Zimmerman [4], [5], about extended binary Goppa codes defined by polynomials G(x) of the following form :

$$G(x) = X^2 + X + \beta \qquad (\beta \in F^*_{2^m})$$

The principal purpose of these authors was to search cyclic Goppa codes : for this, they determine a set of permutations which leaves the code invariant ; one of these permutations is shown to be a cyclic permutation. In [3] O. Moreno uses automorphisms to classify cubic and quartic binary extended Goppa Codes.

In this note, we exhibit a subgroup of the automorphism group of two families of extended Goppa codes over F_q (q odd), which are defined by polynomials of the following form :

$$G_1(X) = (X + \varepsilon\mu)^t \qquad (\varepsilon = +1 \text{ or } \varepsilon = -1)$$
$$G_2(X) = [(X - \mu)(X + \mu)]^t$$

($\mu \in \mathbb{F}^*_{q^m}$, t some power of q)

we then deduce that these codes are <u>cyclic</u> codes.

In first part, we give some definitions and some well-known properties about Goppa Codes. In a second one, we present our results.

PART I

RECALLS [5]

Let : L : $\{\alpha_0, \alpha_1, \ldots, \alpha_{N-1}\}$ a subset of F_{q^m} and let G(x) a polynomial in $F_{q^m}[x]$ with degree $h < N$, such that :

$$G(\alpha_i) \neq 0 \qquad (0 \leq i \leq N-1)$$

Any N-uple will be subscriped by the elements of the set L ; we denote :

$a(\alpha_k)$ for the $(k+1)^{\text{ième}}$ component of a vector a.

For every vector $a = (a(\alpha_0), \ldots, a(\alpha_{N-1}))$ such that $a(\alpha_k) \in \mathbb{F}_q$ $(0 \leq k \leq N-1)$, we consider :

$$R_a(x) = \sum_{k=0}^{N-1} \frac{a(\alpha_k)}{X - \alpha_k} .$$

Definition 1 : The Goppa code $\Gamma(L,G)$ is the set of vectors a with all components in F_q, such that :

$$R_a(x) = 0 \qquad \text{modulo } G(x).$$

Code characteristics : $\Gamma(L,G)$ is a linear code over F_q

- with length $N = |L|$
- with dimension $k \geq N - mh$ (h = degree of G(x))
- with distance $d \geq h+1$.

Parity matrix [5] : $\Gamma(L,G)$ admits a parity matrix H of the following form :

$$H = \begin{bmatrix} 1 & 1 & \cdots & 1 \\ \alpha_0 & \alpha_1 & \cdots & \alpha_{N-1} \\ \alpha_0^2 & \alpha_1^2 & \cdots & \alpha_{N-1}^2 \\ \vdots & \vdots & & \vdots \\ \alpha_0^{h-1} & \alpha_1^{h-1} & \cdots & \alpha_{N-1}^{h-1} \end{bmatrix} \times \begin{bmatrix} G(\alpha_0)^{-1} & & & \text{o} \\ & G(\alpha_1)^{-1} & & \\ & & \ddots & \\ \text{o} & & & G(\alpha_{N-1})^{-1} \end{bmatrix}$$

Therefore, a vector $a = (a(\alpha_0),\dots,a(\alpha_{N-1}))$ belongs to $\Gamma(L,G)$ if and only if we have :

$$\sum_{\alpha_k \in L} \frac{\alpha_k^i \, a(\alpha_k)}{G(\alpha_k)} = 0 \qquad (i = 0,\dots,h-1)$$

Extended code $\hat{\Gamma}(L,G)$

Definition 2 : The extended code $\hat{\Gamma}(L,G)$ of the code $\Gamma(L,G)$ is the code obtained by adding to every code word

$$a = ((\alpha_0),\ \dots,a(\alpha_{N-1})) \text{ in } \Gamma(L,G)$$

the component $a(\infty)$ defined by :

$$a(\infty) = -\sum_{\alpha_k \in L} a(\alpha_k).$$

Property 1 [5] : The vector $a = (a(\alpha_0),\dots,a(\infty))$ belongs to $\hat{\Gamma}(L,G)$ if and only if :

$$\sum_{\alpha_k \in L \cup \{\infty\}} \frac{\alpha_k^i \, a(\alpha_k)}{G(\alpha_k)} = 0 \qquad (i = 0,\dots,h)$$

By convention : $\frac{1}{\infty} = 0$.

PART II

Let (L,G_2) the Goppa code over F_q (q odd) defined by :

$$L = F_{q^m} \setminus \{-\mu \; ; \; \mu\} \qquad (\mu \in \mathbb{F}^*_{q^m}) \; (m > 1)$$

and

$$G_2(x) = [(X - \mu)(X + \mu)]^t$$

$$(\text{deg. of } G_2 = 2q^r < \frac{q^m-1}{m}).$$

We have the following lemmas :

Lemma 1. Let K be the set of mappings π :

$$F_{q^m} \cup \{\infty\} \to F_{q^m} \cup \{\infty\} \; : \; x \to \frac{\eta x + b}{cx + \eta}$$

where η, b, c in F_{q^m} are such that :

$$\eta^2 - bc \neq 0 \text{ and } b = c\mu^2. \text{ Then :}$$

a) K is a subgroup of $PGL_2(F_{q^m})$ with order $|K| = q^m(q^m-2)$.
b) The restriction to $L \cup \{\infty\}$ of $\pi \in K$, is a permutation of $L \cup \{\infty\}$.

Proof : a) We use the matrix representation of π :

$$\pi = \begin{bmatrix} \eta & b \\ c & \eta \end{bmatrix}$$

Note that π and $\lambda\pi$ ($\lambda \in \mathbb{F}^*_{q^m}$) give the same permutation of $\mathbb{F}_{q^m} \cup \{\infty\}$.

Let $\pi = \begin{bmatrix} \eta & b \\ c & \eta \end{bmatrix}$ and $\pi' = \begin{bmatrix} \eta' & b' \\ c' & \eta' \end{bmatrix}$ two elements of the set K. We have :

$$\pi'\pi^{-1} = \frac{1}{\eta^2-bc} \begin{bmatrix} \eta\eta'-b'c & \eta b'-\eta' b \\ \eta c'-\eta' c & \eta\eta'-bc' \end{bmatrix}$$

$$\left.\begin{array}{l} b = c\mu^2 \\ b' = c'\mu^2 \end{array}\right\} \Rightarrow \eta b'-\eta' b = \eta c'\eta^2 - \eta' c\eta^2 = \eta^2(\eta c'-\eta' c)$$

Therefore : $\pi'\pi^{-1} \in K$.
Obviously : identity $\in K$.
So, K is a subgroup of $PGL_2(F_{q^m})$.

The two conditions : $\eta^2 - bc \neq 0$ and $b = c\mu^2$ lead easily to the order of K :

$$|K| = q^m(q^m-2).$$

b) One can easily verify that the two elements μ and $-\mu$ are left fixed under every permutation π in k.
So every π in K can be considered as a permutation of $L \cup \{\infty\}$.

Lemma 2 : The extended Goppa Code $\hat{\Gamma}(L,G_2)$ is invariant under the subgroup K.

Proof : Let $a = (a(0),\ldots,a(\infty))$ a code word in $\hat{\Gamma}(L,G_2)$.

Let $a' = (a'(\beta))_{\beta \in L\cup\{\infty\}}$ the vector obtained from code vector a under some permutation of subgroup K. One can write :

$$a'(\beta) = a\left(\frac{\eta\beta + b}{c\beta + \eta}\right)$$

From property 1, we have to show that :

$$\omega = \sum_{\beta \in L\cup\{\infty\}} \frac{\beta^i a(\beta)}{G_2(\beta)} = 0 \qquad i = 0,\ldots,h \qquad (h = 2q^r)$$

Let : $$\beta' = \frac{\eta\beta + b}{C\beta + \eta}$$

That is : $$\beta = \frac{\eta\beta' - b}{-c' + \eta}$$

We have : $$G_2(\beta) = \frac{(\eta^2 - \eta^2 c^2)q^r}{(\eta - c\beta')^{2q^r}} G_2(\beta')$$

Then : $$\omega = \frac{1}{(\eta^2-\eta^2c^2)q^r} \sum_{\beta' \in L\cup\{\infty\}} \frac{(\eta\beta'-b)^i(\eta-c\beta')^{2q^r-i}a(\beta')}{G_2(\beta')}$$

Let $$P(\beta') = (\eta\beta'-b)^i(\eta-c\beta')^{2q^r-i} = \sum_{j=0}^{2q^r} P_j\beta'^j.$$

Since degree of $P(x) \leq 2q^r = h$, we deduce, from property 1, that

$$\omega = 0.$$

Therefore every permutation $\pi \in K$ leaves invariant the code $\hat{\Gamma}(L,G_2)$.

Now, we exhibit another automorphisms subgroup.

Lemma 3 : The mapping

$\Theta : \mathbb{F}_{q^m} \cup \{\infty\} \to \mathbb{F}_{q^m} \cup \{\infty\} : x \to \mu^{1-q^r} x^{q^r}$ is a permutation of $F_{q^m} \cup \{\infty\}$ with order $\dfrac{m}{GCD(m,r)}$. The extended Goppa Code $\hat{\Gamma}(L,G_2)$ is invariant under the subgroup S generated by Θ.

Proof : Obvioulsy, Θ is a permutation of $F_{q^m} \cup \{\infty\}$.

Since $\Theta^i(x) = \mu^{1-q^{ir}} x^{q^{ir}}$, we deduce that Θ has order $\dfrac{m}{GCD(m,r)}$. Elements μ and $-\mu$ are fixed by Θ, so we can consider Θ as a permutation of $L \cup \{\infty\}$.

Now, let $a = (a(0,\ldots,a(\beta))$ a code word in $\hat{\Gamma}(L,G_2)$.
Let $a' = (a'(0,\ldots,(a'(\beta))$ the vector obtained from code vector a under the permutation Θ. We have :

$$a'(\beta) = a(\mu^{1-q^r}\beta^{q^r}).$$

From property 1, we have to show that :

$$\omega = \sum_{\beta \in L \cup \{\infty\}} \frac{\beta^i a'(\beta)}{G_2(\beta)} = 0 \qquad i = 0,\ldots,2q^r.$$

Let : $\beta' = \mu^{1-q^r}\beta^{q^r}$

That is : $\beta^{q^r} = \beta' \mu^{q^r-1}$.

We have : $\omega = \sum_{\beta} \dfrac{\beta^i a(\beta')}{G_2(\beta)}$

and : $$\omega^{q^r} = \sum_{\beta} \frac{(\beta^{q^r})^i (a(\beta'))^{q^r}}{(G_2(\beta))^{q^r}}$$

It is easy to verify that :

$$[G_2(\beta)]^{q^r} = \mu^{2(q^r-1)q^r}\ G_2(\beta').$$

On the other hand, since $a(\beta')$ belongs to F_q, we have

$$[a(\beta')]^{q^r} = a(\beta')$$

So :

$$\omega^{q^r} = \frac{(\mu^{q^r-1})^i}{\mu^{2(q^r-1)q^r}} \sum_{\beta' \in L\cup\{\infty\}} \frac{\beta'^i a(\beta')}{G_2(\beta')} \qquad i = 0,\ldots,h$$

From property 1, we deduce that

$$\omega^{q^r} = 0 \qquad i = 0,\ldots,h = 2q^r$$

Hence : $\underline{\omega = 0}$.

That prooves that $\hat{\Gamma}(L,G_2)$ is left invariant by Θ.

Finally, we have the following result :

<u>Proposition 1</u> : Let $\hat{\Gamma}(L,G_2)$ be the extended Goppa Code defined by

$$L = F_{q^m} \setminus \{\mu\ ;\ -\mu\} \qquad (\mu \in F^*_{q^m})$$

and : $G_2(X) = [(X-\mu)(X+\mu)]^{q^r}$ $\qquad (q^r < \frac{q^m-1}{2m})$

Let S the subgroup generated by $\Theta : x \to \mu^{1-q^r} x^{q^r}$ and let K the subgroup consisting of the permutations $\pi(\eta,b,c)$:

$$x \to \frac{\eta x + b}{cx + \eta} \qquad (\eta,b,c \text{ in } F_{q^m} \text{ such that : } \eta^2 - bc \neq 0,\ b = c\mu^2)$$

Then :

a) SK is a subgroup with order $|S| \times |K|$, which leaves invariant the code $\hat{\Gamma}(L,G_2)$.

b) $\hat{\Gamma}(L,G_2)$ is a cyclic code.

<u>Proof</u> : a) If remains to proove that SK is a subgroup :

Let $\Theta \in S$ and $\pi(\eta,b,c) \in K$.

Choose : $\eta' = \eta^{q^r}$
$b' = \eta^{1-q^r} b^{q^r}$
and $c' = \eta^{q^r-1} c^{q^r}$.

We have : $b' = c'\eta^2$, so $\pi(\eta',b',c') \in K$.

It is easy to verify the following relation :

$$\Theta \circ \pi(\eta,b,c) = \pi(\eta',b',c') \circ \Theta$$

So : $SK = KS$.

We can conclude : SK is a subgroup of the full automorphisms group of $\hat{\Gamma}(L,G_2)$.

Since $S \cap K = \{\text{Identity}\}$, we have $|SK| = |S||K|$.

Remark : Law in SK is a semi-direct product Law.

b) To show that $\hat{\Gamma}(L,G_2)$ is a cyclic code, we proove the existence of an element π in K of order q^m-1.

$$\text{Let } \pi = \begin{bmatrix} \eta & b \\ 1 & \eta \end{bmatrix} \in K.$$

(Recall the π operates over $\mathbb{F}_{q^m} \cup \{\infty\}$ when we identify vector $\begin{bmatrix} r \\ s \end{bmatrix}$ with element $\frac{r}{s}$ in $\mathbb{F}_{q^m}$ when $s \neq 0$, and vector $\begin{bmatrix} 1 \\ 0 \end{bmatrix}$ with point ∞).

The eigenvalues of π are :

$$\lambda_1 = \eta-\mu \quad \text{and} \quad \lambda_2 = \eta+\mu$$

So matrix $\begin{bmatrix} \eta & b \\ 1 & \eta \end{bmatrix}$ is equivalent to matrix $\begin{bmatrix} \lambda_1 & 0 \\ 0 & \lambda_2 \end{bmatrix}$.

Like η, $\eta-\mu$ takes q^m-2 distinct values in $\mathbb{F}_{q^m}$. Since the number of primitive elements in F_{q^m} is such that

$$\varphi(q^m-1) > 2$$

we deduce that there exists an element η_0 and a primitive element α in $\mathbb{F}_{q^m}$ such that

$$\alpha = \eta_0 - \mu.$$

Therefore the order of $\begin{bmatrix} \lambda_1 & 0 \\ 0 & \lambda_2 \end{bmatrix}$ is q^m-1, so is the order of $\pi(\eta_0,b,1)$.

The second family.

The proofs are quite similar. We only give the result (also quite similar to case 1).

Proposition 2 : Let $\hat{\Gamma}(L,G_1)$ be the extended Goppa code defined by :

$$L = F_{q^m} \setminus \{-\mu ; \mu\} \qquad \mu \in F^*_{q^m}$$

and

$$G_1(X) = (X + \varepsilon\mu)^{q^r} \qquad q^r < \frac{q^m-1}{m} \qquad \varepsilon = +1 \text{ or } \varepsilon = -1$$

Then :

a) $\hat{\Gamma}(L,G_1)$ is left invariant by the subgroup SK of the full automorphism group, where

- S is the subgroup generated by Θ :

$$\Theta : x \to \mu^{1-q^r} x^{q^r}.$$

- K is the subgroup consisting of the permutations

$$\pi(\eta,b,c) : x \to \frac{\eta x + b}{cx + \eta}$$

η,b,c in F_{q^m} such that $\eta^2-bc \neq 0$ and $b = c\mu^2$. The order of SK is :

$$|SK| = \frac{m}{GCD(m,r)} \times q^m(q^m-2).$$

b) $\hat{\Gamma}(L,G_1)$ is a cyclic code.

Conclusion

In this note, we have described a subgroup of the full automorphisms group of two particular extended Goppa codes over F_q (q odd).

Now, a natural purpose is to perform a decoding procedure of these codes by using such a subgroup.

References :

[1] - E.R. Berlekamp, O. Moreno
"Extended Double Error Correcting Binary Goppa Codes are Cyclic". IEEE, nov. 1973. p. 817-818.

[2] - V.D. Goppa : "A new class of Linear Error Correcting Codes". Prob. Peredach Inform, Vol. 6, pp. 24-30.

[3] - O. Moreno : "Symmetries of binary Goppa Codes". IEEE Transactions on Information Theory, Vol. TT 25, n° 9, p. 609-612, Sept. 1979.

[4] - K.K. Tzeng, K. Zimmerman : "On Extending Goppa Codes to Cyclic Codes". IEEE, nov. 1975, p. 712-716.

[5] - F.J. Mc Williams, NJA Sloane : "The Theory of Error Correcting Codes". North-Holland, 1977.

SOME QUASI-PERFECT CYCLIC CODES

J.L DORNSTETTER

LABORATOIRE CENTRAL DE TELECOMMUNICATIONS

18-20 Rue Grange Dame Rose

78141 VELIZY-VILLACOUBLAY CEDEX FRANCE

Tel.(33) 1 39 46 96 15

We show that the Zetterberg and Melas codes are quasi-perfect

I.

We first prove that the Zetterberg codes are quasi-perfect . The proof for the Melas codes requires minor changes that are given in section II .

Consider the cyclic code C of length $n = 2^m+1$, m even , whose generator polynomial g(X) is the minimal polynomial of some primitive n^{th} root of unity in the field $GF(2^{2m})$.

Since $2^m = -1$ (mod n) , we have $g(a)=0 \Rightarrow g(a^{-1})=0$ and C is reversible . Its dimension is $n-2m = 2^m+1-2m$.

<u>LEMMA 1</u> : The minimum distance of C is 5 .

Various proofs of this are known , however we shall present a very simple one .

Proof :

The BCH bound gives $d > 2$; (In fact , a clever use of the BCH bound yields $d > 3$).

Suppose $d = 3$. Consider the syndrome associated with a given error pattern as an element of $GF(2^{2m})$; This is legitimate since $GF(2^{2m})$ is isomorphic to the set of residue classes mod $g(X)$.

For a single error in position i , the syndrome S is given by

$$S = a^i \quad .$$

We have $S^n = 1$ for every syndrome corresponding to a single error .

For two errors in positions i and j , $S = a^i + a^j$. Assuming $d = 3$, there would exist i,j such that $(a^i + a^j)^n = 1$; Setting $r = a^{i-j}$ we would have $r + r^{-1} = 1$; But this implies $r^3 = 1$, in contradiction with the fact that $(n,3) = 1$ so that $d > 3$.

Suppose now $d = 4$. A minimum weight codeword would belong to the cyclic code generated by $(X+1)g(X)$ whose minimum distance is at least 6 by the BCH bound . Thus we have $d > 4$; In fact $d = 5$ since otherwise , the shortened code $(n-1,k,d-1)$ would violate the volume bound . Q.E.D.

Next , we give a simple caracterization of the cosets that admits a leader of weight 2 , which enables us to show by a counting argument that C is a quasi-perfect code .

LEMMA 2 : A coset with syndrome S admits a leader of weight two iff $T(S^{-n}) = 1$.

$T(\ )$ is the usual trace from $GF(2^m)$ to $GF(2)$.

Proof :

" Only if " :

Assume 2 errors in positions i and j . Then $S = a^i + a^j$ and $S^n = S^{n-1}.S = (a^{2^m.i} + a^{2^m.j}).(a^i + a^j) = (a^{-i} + a^{-j}).(a^i + a^j)$. Setting $r = a^{i-j}$ $(r \neq 1)$ we obtain $S^n = r + r^{-1}$.

The following quadratic

$$X^2 + S^n.X + 1 = 0$$

admits two roots , namely r and r^{-1} that are in $GF(2^{2m})$ but not in $GF(2^m)$. It is therefore irreducible over $GF(2^m)$ and $T(S^{-n}) = 1$.

" If " :

The minimum distance of C is 5 by Lemma 1 . We then have : all syndromes corresponding to error patterns of weight 2 are distinct . Their number is $n(n-1)/2$ and this is precisely the number of elements in $GF(2^{2m})$ satisfying the condition $T(S^{-n}) = 1$. Q.E.D.

THEOREM 1 : The Zetterberg codes are quasi-perfect .

Proof :

We have to show that every coset H with syndrome S in $GF(2^{2m})$ admits a leader of weight 0,1,2 or 3 . Taking into account

Lemmas 1 and 2 , the only remaining case is $T(S^{-n}) = 0$, $S^n \neq 0 , 1$. We shall prove that such a coset always admits a leader of weight 3 . The idea is to show that it is possible to change one bit in the noisy codeword so that the new syndrome satisfyes the conditions of Lemma 2 . Since C is cyclic , we can use a cyclic shift so as to insure that $S \in GF(2^m)$; (Multiply S by an appropriate power of a) . The syndrome S_k of the coset H_k that differs from H in the kth bit is

$$S_k = S + a^k .$$

We compute : $S_k^n = (S + a^k)^{n-1} (S + a^k) = S^2 + 1 + S (a^k + a^{-k})$. With k considered mod n , we have $S_k^n = S_{-k}^n$ and S_k^n assumes exactly $(n+1)/2$ distinct values in $GF(2^m)$, all different from 0 and 1 . The pigeon-hole principle , together with the fact that there exits only $(n-5)/2$ elements in $GF(2^m)$ distinct from 0 and 1 such that $T(X^{-1}) = 0$, implies that there exists at least 5 values of k such that $T(S_k^{-n}) = 1$. This proves that H admits several coset leaders of weight 3 . Q.E.D.

The proof of the theorem yields a complete decoding algorithm . The only anoying case is $T(S^{-n}) = 0$, $S^n \neq 0,1$. We first multiply S by a suitable n^{th} root of unity and then search for an $X_k = a^k + a^{-k}$ such that $T ((S^2 + 1 + S.X_k)^{-1}) = 1$.

The sequence X_k can be easily generated as follows :

$$X_0 = 0 , X_1 = a + a^{-1} , X_{k+1} = X_1 . X_k + X_{k-1} \quad \text{for } k > 0 .$$

We were unable to prove that this procedure terminates quickly (roughly in m iterations , say) although there is a considerable experimental evidence that it does . It happens that for $m = 4,6,8$ there always exists a good value of k in the set $1,2,4,8,..,2^m$. For such m , the search can be performed by successive squaring of X or S. It would be interesting to know if this remains true for larger m .

Similar results are now presented for the Melas codes .

II. The definition of the Melas codes is similar to that of the Zetterberg codes ; We consider now the cyclic code C' of length $n = 2^m-1$, m odd , whose generator polynomial g(X) is given by

$$g(X) = M_1(X).M_{-1}(X)$$

M_1 is the minimal polynomial of a , an nth root of unity in $GF(2^m)$, $M_{-1}(X) = X^m M_1(X^{-1})$. Like in the previous case , C' is reversible and its dimension is equal to 2^m-1-2m .

<u>LEMMA 3</u> : The minimum distance of C' is 5 .

Proof:

The BCH bound gives $d > 2$.

The syndrome of a noisy codeword (polynomial) R is $\{ R(a),R(a^{-1}) \} = \{ S_+, S_- \}$. S_+ and S_- are in $GF(2^m)$. We note $S = S_+ . S_-$

For a single error in position i , $S_+ = a^i$, $S_- = a^{-i}$, so that $S = 1$.

For two errors in position i and j , $S_+ = a^i + a^j$, $S_- = a^{-i} + a^{-j}$.

If $d = 3$, there exist i and j such that $(a^i + a^j).(a^{-i} + a^{-j}) = 1$.

Setting $r = a^{i-j}$, this implies $r^3 = 1$, contradicting $(n,3) = 1$.

As in Lemma 1 , d=4 is impossible and $d = 5$. Q.E.D.

<u>LEMMA 4</u> : A coset with syndrome $\{ S_+, S_- \}$, $S \neq 0$ admits a leader of weight two iff $T((S_+.S_-)^{-1}) = 0$

Proof : " Only if " :

As in Lemma 2 , for two errors in position i and j , the quadratic

$$X^2 + S.X + 1 = 0$$

admits two distinct roots in $GF(2^m)$, namely r and r^{-1} , $r = a^{i-j}$, so that $T(S^{-1}) = 0$ holds .

" If " :

The counting argument is identical to that used in the proof of Lemma 2 but here the number of pairs $\{ S_+, S_- \}$ such that $S_+.S_- \neq 0$ and $T((S_+.S_-)^{-1}) = 0$ is equal to $(2^m-1)(2^{m-1}-1) = n(n-1)/2$. Q.E.D.

<u>Theorem</u> : The Melas codes are quasi-perfect .

Proof :

There remains two cases not covered by Lemmas 3 and 4 . We begin with the case $S = 0$ but , say , $S_- \neq 0$.

Consider a coset of C′ containing a minimum weight codeword W of the Hamming code with generator polynomial $M_1(X)$. The corresponding syndrome is such that $S'_+ = 0$, $S'_- \neq 0$.

There exists a cyclic shift of W that yields $S'_- = S_-$, proving the existence of a coset leader of weight 3 . (Of course , a similar argument works if we exchange S_- and S_+) .

The last case is $S \neq 1$, $T(S^{-1}) = 1$.

As for the Zetterberg codes , we use a cyclic shift to insure $S_+ = S_-$ and we then try to invert the kth bit so as to obtain a modified syndrome S'_k satisfying the condition of Lemma 4 :

$$S'_k = S_+.S_- + 1 + S_+(a^k + a^{-k})$$

assumes exactly (n+1)/2 distinct values , none of which is 1 (this would imply the existence of a coset leader of weight 2) while there exists only (n-1)/2 elements in $GF(2^m) - \{ 1 \}$ such that $T(X^{-1}) = 1$. This proves that there exists a " good choice " for k . Q.E.D.

Dr G.Cohen has drawn our attention to (ref.[1]) that contains a proof of the above theorems . However , Moreno's proof requires some non-trivial knowledge about the Zetterberg and Melas codes , namely the fact that they can be derived from a Goppa code G and the knowledge of the number of minimum weight codewords in G .(The proof also relies on counting arguments) .

On the other hand , it yields more results since it also shows the Goppa codes $G(2^{2m+1}, 2^{2m+1}-4m-2, 5)$ are quasi perfect .

Aknowledgment.

We would like to thank Prof. O.Moreno for pointing out that the proof originally presented for the Zetterberg codes also applied for the Melas codes .

References .

[1] : O.Moreno : " Quasi-perfect double error correcting codes related to Goppa codes " . Preprint , to appear in IEEE Trans. on Inf.Theory .

Explicit Kerdock Codes over GF(2)

D.A. Leonard and C.A. Rodger
Auburn University, Auburn, Alabama 36849 U.S.A.

This paper gives a quick, easy method for producing (and hence encoding) a Kerdock code, given little more than a primitive polynomial of odd degree over GF(2) (and the corresponding index table.) It is hoped that this will enable someone to produce an equally quick, easy decoding scheme to enrich the literature on interesting non-linear codes.

van Lint [5], in his invited lectures at the fourteenth Southeastern Conference on Combinatorics, Graph Theory, and Computing, gave a very interesting talk on the formally dual non-linear Preparata codes and Kerdock codes, and bemoaned the limited number of readable papers in the literature on non-linear codes with interesting structure.

His exposition on Preparata codes, based on a paper by Baker, van Lint, and Wilson [1], gave a description of those codes from which it is relatively easy to produce encoding and decoding algorithms using linear techniques similar to those used in 2-error correcting BCH codes. However the description of Kerdock codes, based on work by Kantor [4], Dillon [2], and Dye [3], was still of a highly geometric nature, and not so easy to wade through.

The method proposed here to produce a Kerdock code over GF(2) hides the geometry, and uses easy finite field calculations instead.

A Kerdock code $K(2m)$ is gotten by carefully choosing 2^{2m-1} cosets of the $2^{\binom{m}{2}}$ cosets (of the first-order Reed-Muller code $RM(1,2m)$) that make up the second-order Reed-Muller code $RM(2,2m)$ in order to get a non-linear code of word length 2^{2m}, with $2^{2m+1} \cdot 2^{2m-1} = 2^{4m}$ codewords, and a minimum distance of $2^{2m-1} - 2^{m-1}$.

The procedure alluded to by van Lint was to somehow use

1) the trace function $tr: GF(2^{2m-1}) \to GF(2)$ to produce

2) a spread of $(2m-1)$-dimensional subspaces of $(4m-2)$-dimensional space, extend this somehow to

3) a spread of $(2m)$-dimensional totally isotropic subspaces of type I for the quadric

$$Q = \{ x : \sum_{i=0}^{2m-1} x_i \, x_{2m+i} = 0 \}$$

in (4m)-dimensional space, and read from that

4) a Kerdock set of $2m \times 2m$ non-singular, symmetric matrices with zero diagonal such that the sum of any two is also non-singular, which easily gives

5) a Kerdock set of homogeneous quadratic forms that describe the cosets of RM(1,2m) needed to make up K(2m).

We suggest a simpler method of producing a Kerdock set by

A) using any primitive polynomial of degree $2m - 1$ over GF(2) to produce

B) an ordered $(2m-1)$ tuple of "seed" powers of the generator of the corresponding finite field, to get

C) a pre-Kerdock set of $(2m-1) \times (2m-1)$ nonsingular, symmetric matrices, such that the sum of any two is also non-singular, then "excising" the diagonals to get a Kerdock set of $2m \times 2m$ matrices as above.

It may be helpful to the reader to follow the example at the end of the paper, in which the eight coset representatives for K(4), the Nordstrom-Robinson code, are constructed using this method.

Let $p(x) = x^{2m-1} + \sum_{i=0}^{2m-3} a_i x^i$ be a primitive polynomial over GF(2), so the $K = GF(2)\,[\alpha]/p(\alpha) \simeq GF(2^{2m-1})$ and α is a generator for the multiplicative group of K.

Define $y_j = \alpha^{t_j} = \begin{cases} \sum_{i=0}^{2m-3} a_{i+j-(2m-3)} \alpha^i, & 0 \le j \le 2m-3 \\ \alpha^{2m-2}, & j = 2m-2 \end{cases}$

and $y_{k,j} = \alpha^k y_j = \sum_{i=0}^{2m-2} A_{k,i,j} \alpha^i$. Let A_k be the matrix with (i,j)th entry $A_{k,i,j}$.

<u>Proposition 1</u>. A_k is symmetric, $1 \le k \le n = 2^{2m-1} - 1$

<u>Proof</u>. A_0 is symmetric by definition. Recursively the jth row of A_{k+1} is α times the jth row of A_k, whereas the jth column of A_{k+1} is the (j-1)st column of A_k plus a_j times the (2m-2)nd column of A_k. But

$$\alpha \, y_{k,j} = \alpha^k \, (\alpha y_j) = \alpha^k (y_{j-1} + a_j y_{2m-2}) = y_{k,j-1} + a_j y_{k,2m-2}$$

<u>Proposition 2</u>. If $\sum_{k=1}^{n} b_k \alpha^k \neq 0$, then $A = \sum_{k=1}^{n} b_k A_k$ is non-singular.

Proof. Suppose $\underline{c}A = \underline{0}$. Then

$$0 = \sum_{j=0}^{2m-2} c_j \sum_{k=1}^{n} b_k \alpha^{k+t_j} = \sum_{j=0}^{2m-2} c_j \alpha^{t_j} \sum_{k=1}^{n} b_k \alpha^k .$$

So $0 = \sum_{j=0}^{2m-2} c_j \alpha^{t_j}$, and $\underline{c} = \underline{0}$.

Corollary 3. $A_k - 0$ and $A_k - A_\ell$ are non-singular for $k \neq \ell$, $1 \leq k,\ell \leq 2^{2m-1} - 1$.

For any square matrix A, define the diagonal vector $\underline{d}_A$ by $(\underline{d}_A)_i = A_{i,i}$.

Proposition 4. If A is a symmetric matrix over $GF(2)$, then $\underline{x}A\underline{x}^t = \underline{d}_A\underline{x}^t$.

Proof. $\underline{x}A\underline{x}^t = \sum_{i,j} x_i A_{i,j} x_j = \sum_i A_{i,i} x_i^2 + \sum_{i<j} (A_{i,j} + A_{j,i}) x_i x_j$

$$= \sum_i A_{i,i} x_i = \underline{d}_A \underline{x}^t.$$

Proposition 5. If A is symmetric over $GF(2)$ and non-singular, then $\underline{d}_A A^{-1} = \underline{d}_{A^{-1}}$.

Proof. Using prop. 4 with $\underline{x}^t$ the jth column of A^{-1}, $\underline{d}_A A^{-1} = \underline{d}_{(A^{-1})^t A A^{-1}} = \underline{d}_{(A^{-1})^t} = \underline{d}_{A^{-1}}$.

Proposition 6. If A is a symmetric $m \times m$ matrix over $GF(2)$, and non-singular, then $\underline{d}_A \underline{d}_{A^{-1}}^t \equiv n \pmod 2$.

Proof. Given any symmetric matrix A over $GF(2)$ there is a non-singular matrix S so that $S^t AS$ is zero except for diagonal blocks of size 1 or 2 and such that $S^t AS$ and $S^{-1}A^{-1}(S^{-1})^t$ agree on the diagonal only on the blocks of size 1.

But $(S^t AS)_{i,i} = \sum_{j,k} S_{j,i} A_{j,k} S_{k,i} = \sum_j A_{j,j} S_{j,i}^2$

$+ \sum_{j<k} (A_{j,k} + A_{k,j}) S_{j,i} S_{k,i} = \sum_j A_{j,j} S_{j,i}$

So $\underline{d}_{S^t AS} = \underline{d}_A S$. But then $\underline{d}_A \underline{d}_{A^{-1}}^t = \underline{d}_A SS^{-1} \underline{d}_{A^{-1}}^t = (\underline{d}_A S)(\underline{d}_{A^{-1}}(S^{-1})^t)$

$= \underline{d}_{S^t AS} \underline{d}_{S^{-1}A^{-1}(S^{-1})^t} \equiv n \pmod 2$

Proposition 7. If A is a symmetric, non-singular $(2m-1) \times (2m-1)$ matrix over $GF(2)$, then

$$\overline{A} = \left(\begin{array}{c|c} A + \underline{d}^t{}_A\,\underline{d}_A & \underline{d}^t{}_A \\ \hline \underline{d}_A & 0 \end{array}\right)$$

is non-singular.

<u>Proof</u>. If $(\underline{x},z)\overline{A} = (\underline{0},0)$. then

$\underline{x}A + \underline{xd}^t{}_A\underline{d}_A + z\underline{d}_A = \underline{0}$ and $\underline{xd}^t{}_A = 0$

So $\underline{x}A = z\underline{d}_A$, and hence $\underline{x} = z\underline{d}_A A^{-1} = z\underline{d}_{A^{-1}}$

But then $\underline{xd}_A{}^t = z\underline{d}_{A^{-1}}\underline{d}_A{}^t = 1$.

<u>Proposition 8</u>. If A and B are symmetric, non-singular $(2m-1) \times (2m-1)$ matrices over $GF(2)$ and $A + B$ is non-singular then $\overline{A} + \overline{B}$ is non-singular, if $\underline{d}_A \neq \underline{d}_B$.

<u>Proof</u>. If $(\underline{x},z)(\overline{A} + \overline{B}) = (\underline{0},0)$, then

$\underline{x}(A + B) = \underline{xd}_A{}^t\,\underline{d}_A + \underline{xd}_B{}^t\,\underline{d}_B + z(\underline{d}_A + \underline{d}_B) = \underline{0}$

and $\underline{x}(\underline{d}_A{}^t + \underline{d}_B{}^t) = 0$. So $\underline{x}(A + B) = (\underline{xd}_A{}^t + z)(\underline{d}_A + \underline{d}_B) \neq 0$.

So $\underline{xd}_A{}^t + z = 1$ and $\underline{x} = \underline{d}_{(A+B)}(A + B)^{-1} = \underline{d}_{(A+B)^{-1}}$

But then $\underline{xd}_{A+B}{}^t = \underline{d}_{(A+B)^{-1}}\underline{d}_{A+B}{}^t = 1$.

Define $V_\infty = \{(0, \ldots, 0; X_0, \ldots, X_{2m-2}); x_i \in GF(2)\}$

$V_0 = \{(x_0, \ldots, x_{2m-2}; 0, \ldots, 0): x_i \in GF(2)\}$

$V_\alpha k = \{(x_0, \ldots, x_{2m-2}; z_{k,0}, \ldots, z_{k,2m-2}); x_i \in GF(2)\}$

Where $z_{k,j} = \sum_{i=0}^{2m-2} A_{k,i,j}x_i$

<u>Proposition 9</u>. V_∞, V_0, $V_\alpha k$, $A \leq k \leq 2^{2m-1} - 1$ is a spread of $(2m-1)$-dimensional subspaces of $4m-2$ space.

<u>Proof</u>. $V_\infty \quad V_0 = V_\infty \quad V_\alpha k = \{\underline{0}\}$. If $V_0 \quad V_\alpha k \neq \{\underline{0}\}$ then $z_{k,0} = \ldots = z_{k,2m-2} = 0$, so $\underline{x}A_k = \underline{0}$, but A_k is non-singular. If $V_\alpha k \quad V_\alpha \ell \neq \underline{0}$, then $Z_{k,0} + z_{\ell,0} = \ldots = Z_{k,2m-2} + Z_{\ell,2m-2} = 0$, so $\underline{x}(A_k + A_\ell) = \underline{0}$, but $A_k + A_\ell$ is non-singular. So $V_\infty, V_0, \{V_\alpha k\}_k$ cover $(2^{2m-1}+1)(2^{2m-1}-1) = 2^{4m-2}-1$ non-zero vectors.

Define $\overline{V}_\infty = \{(0, \ldots, 0; x_0, \ldots, x_{2m-1}): x_i \in GF(2)\}$

$\overline{V}_0 = \{(x_0, \ldots, x_{2m-1}; 0, \ldots, 0): x_i \in GF(2)\}$

$\overline{V}_\alpha k = \{(x_0, \ldots, x_{2m-1}; \overline{z}_{k,0}, \ldots, \overline{z}_{k,2m-1}): x_i \in GF(2)\}$

where
$$\overline{z}_{k,j} = \begin{cases} \sum_{i=0}^{2m-2} A_{k,i,j}\, x_i\,, & j = 2m-1 \\ A_{k,i,i,}x_{2m-1} + \sum_{i=0}^{2m-1}(A_{k,i,j} + A_{k,i,j}A_{k,j,j})x_i, & 0 < j < 2m-2 \end{cases}$$

Proposition 10. ∇_∞, ∇_0, $\{\nabla_\alpha k\}_k$ is a spread of (2m-1)-dimensional subspaces of the quadric

$$Q = \{(x_0, \ldots, x_{4m-1}) : \sum_{i=0}^{2m-1} x_i x_{2m+i} = 0\}$$

in 4m-space.

Proof. As for prop. 9 except
$1 + (2^{2m} - 1)(2^{2m-1} + 1) = 2^{4m-1} + 2^{2m-1} = |Q|$.

Example. To find the coset representatives for K(4), begin with the primitive polynomial $p(x) = x^3 + (1+x)$ in GF(2). This gives the seed matrix $A_0 = \begin{pmatrix} 010 \\ 110 \\ 001 \end{pmatrix}$, from which one gets $A_1 = \begin{pmatrix} 001 \\ 011 \\ 110 \end{pmatrix}$,

$A_2 = \begin{pmatrix} 110 \\ 111 \\ 011 \end{pmatrix}$, $A_3 = \begin{pmatrix} 011 \\ 101 \\ 111 \end{pmatrix}$, $A_4 = \begin{pmatrix} 111 \\ 100 \\ 101 \end{pmatrix}$, $A_5 = \begin{pmatrix} 101 \\ 010 \\ 100 \end{pmatrix}$, and

$A_6 = \begin{pmatrix} 100 \\ 001 \\ 010 \end{pmatrix}$, by multiplying the row vectors by the generators of GF(8) and using the index table. Proposition 7 gives the new matrices

$$\bar{A}_0 = \begin{pmatrix} 0100 \\ 1011 \\ 0101 \\ 0110 \end{pmatrix}, \bar{A}_1 = \begin{pmatrix} 0010 \\ 0011 \\ 1100 \\ 0100 \end{pmatrix}, \bar{A}_2 = \begin{pmatrix} 0011 \\ 0001 \\ 1001 \\ 1110 \end{pmatrix}, \bar{A}_3 = \begin{pmatrix} 0110 \\ 1010 \\ 1101 \\ 0010 \end{pmatrix},$$

$$\bar{A}_4 = \begin{pmatrix} 0101 \\ 1000 \\ 0001 \\ 1010 \end{pmatrix}, \bar{A}_5 = \begin{pmatrix} 0111 \\ 1001 \\ 1000 \\ 1100 \end{pmatrix}, \bar{A}_6 = \begin{pmatrix} 0001 \\ 0010 \\ 0100 \\ 1000 \end{pmatrix}.$$

The coset representatives relative to RM(1,4) can then be read off from the entries above the diagonals, namely

$$\underline{f}_0 = \underline{v}_1\underline{v}_2 + \underline{v}_2\underline{v}_3 + \underline{v}_2\underline{v}_4 + \underline{v}_3\underline{v}_4$$
$$\underline{f}_1 = \underline{v}_1\underline{v}_3 + \underline{v}_2\underline{v}_3 + \underline{v}_2\underline{v}_4$$
$$\underline{f}_2 = \underline{v}_1\underline{v}_3 + \underline{v}_1\underline{v}_4 + \underline{v}_2\underline{v}_4 + \underline{v}_3\underline{v}_4$$
$$\underline{f}_3 = \underline{v}_1\underline{v}_2 + \underline{v}_1\underline{v}_3 + \underline{v}_2\underline{v}_3 + \underline{v}_3\underline{v}_4$$
$$\underline{f}_4 = \underline{v}_1\underline{v}_2 + \underline{v}_1\underline{v}_4 + \underline{v}_3\underline{v}_4$$
$$\underline{f}_5 = \underline{v}_1\underline{v}_2 + \underline{v}_1\underline{v}_3 + \underline{v}_1\underline{v}_4 + \underline{v}_2\underline{v}_4$$
$$\underline{f}_6 = \underline{v}_1\underline{v}_4 + \underline{v}_2\underline{v}_3$$
$$\underline{f}_7 = \underline{0}$$

References

[1] R. D. Baker, J. H. van Lint and R. M. Wilson, On the Preparata and Goethals codes, IEEE Trans. Inf. Theory (1983).

[2] J. F. Dillon, On Pall partitions for quadratic forms (1974).

[3] R. H. Dye, Partitions and their stabilizers for line complexes and quadrics, Annali Mat. (4), 114 (1977), 173-194.

[4] W. M. Kantor, Spreads, translation planes and Kerdock sets I, II, SIAM J. Alg. Disc. Math. 3(1982), 151-156 and 308-318.

[5] J. H. van Lint, Kerdock codes and Preparata codes, Congressus Numerantium, Vol. 39 (1983), pp. 25-41.

UNE CLASSE DE CODES 2-CORRECTEURS ADAPTES AUX SYSTEMES D'INFORMATION FORMATES

Courteau Bernard * et Goulet Jean *

Département de Mathématiques et d'Informatique
Université de Sherbrooke
SHERBROOKE (Québec)
Canada J1K 2R1

Introduction

Dans ce travail nous nous intéressons aux systèmes d'information "formatés" où l'information est naturellement décomposée en sous-blocs $b_0, b_1, \ldots, b_{m-1}$ de même longueur ℓ et de même parité. De tels systèmes se présentent par exemple dans les systèmes vidéotex utilisant le code ASCII à 8 bits [1,2], dans les mémoires d'ordinateurs [3] ou dans les systèmes d'enregistrement de données sur bande ou sur disque utilisant un algorithme de compression [4,5].

Pour protéger ces systèmes d'information formatés contre les erreurs aléatoires, nous proposons une classe infinie de codes 2-correcteurs sur un alphabet quelconque GF(q), à taux élevé, de longueur multiple de ℓ et pouvant corriger une grande partie des erreurs triples. Les mots d'un tel code auront la forme

$$[b_0, b_1, \ldots, b_{m-1}, b_m, b_{m+1}] \tag{0}$$

où les b_i sont des suites (appelées sous-blocs ou "bytes") de ℓ éléments de GF(q), les sous-blocs $b_0, \ldots, b_{m-1}$, tous de même parité, étant appelés sous-blocs d'information alors que les deux derniers sous-blocs b_m, b_{m+1} sont appelés sous-blocs de parité.

La construction de ces codes utilise une conique non-dégénérée dans un plan projectif et donne une amélioration du code considéré dans [2] (cas particulier ℓ=8).

1. - Construction

On démontre facilement le théorème suivant dont l'idée essentielle se trouve dans [2].

---

* Les auteurs ont bénéficié des octrois de recherches FCAC# EQ1886 et CRSNG# A5120.

Théorème 1

Soit ℓ et m des entiers naturels et C le $(\ell(m+2), (\ell-1)m)$ -code sur le corps GF(q) admettant une matrice de contrôle H du type

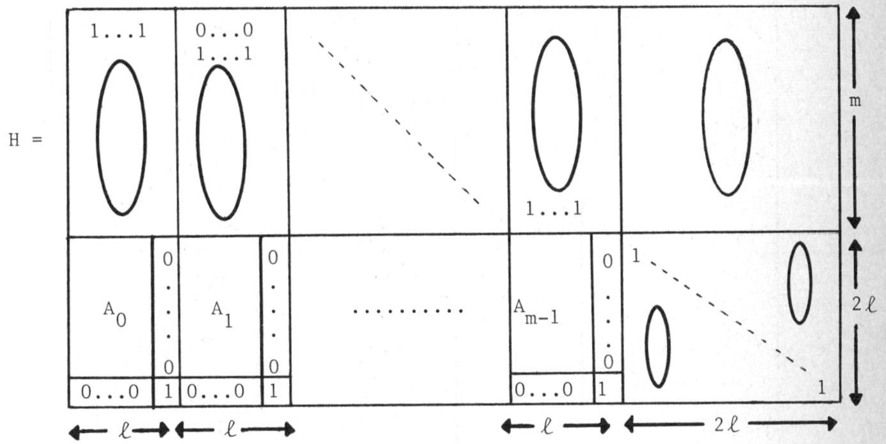

(1)

Si 1° les colonnes de A_i ne sont jamais parallèles,

2° une fois fixée la somme des coefficients, toutes les combinaisons linéaires de deux colonnes de A_i sont différentes,

3° il existe m sous-espaces W_i de $[GF(q)]^{\ell-1}$ tels que les colonnes de A_i sont contenues dans W_i et tels que toute somme de trois sous-espaces distincts $W_{i_1} + W_{i_2} + W_{i_3}$ est directe,

4° le poids minimum de W_{i_2} est ≥ 3,

alors C est un code 2-correcteur et peut, de plus, corriger tous les schémas d'erreur consistant en une erreur dans 3 sous-blocs d'information différents.

Démonstration :

Il suffit de s'assurer que les syndrômes des schémas d'erreurs annoncés dans le théorème sont tous différents.

Nous allons maintenant donner une construction explicite de matrices A_i satisfaisant aux conditions du théorème 1.

Soit q une puissance d'un nombre premier p, ℓ un nombre naturel tel que $a = \frac{2\ell-1}{3}$ soit un entier (i.e ℓ=3k+2 ; ℓ = 2,5,8,11,14,...), α un élément primitif du corps $GF(q^a)$. Désignons par la même lettre β l'image q-aire de $\beta \in GF(q^a)$ relativement à la base naturelle $\{1,\alpha,\dots,\alpha^{a-1}\}$ i.e le vecteur-colonne formé des composantes de β dans cette base.

Formons la matrice quasi-diagonale N_0 de taille $\ell-1=3k+1$ par $a = 2k+1$

$N_0 =$

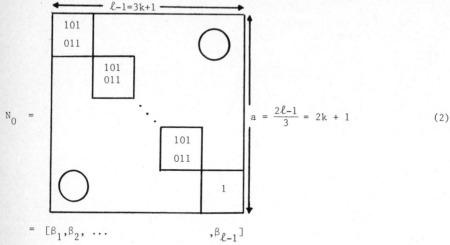

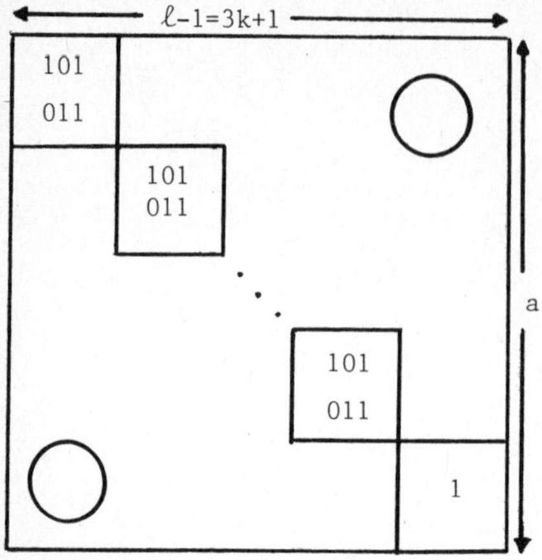

$$a = \frac{2\ell-1}{3} = 2k+1 \qquad (2)$$

$= [\beta_1, \beta_2, \dots \qquad , \beta_{\ell-1}]$

où $\beta_j \in GF(q^a)$ est l'élément dont l'image q-aire est la j^e colonne de N_0.

Considérons les q^a matrices partitionnées

$$A_i = \begin{bmatrix} N_0 \\ \alpha^i N_0 \\ \alpha^{-i} N_0 \end{bmatrix} , \quad i = 0, 1, \dots, q^a-1 \qquad (3)$$

où $\alpha^i N_0 = \alpha^i[\beta_1, \dots, \beta_{\ell-1}] = [\alpha^i\beta_1, \dots, \alpha^i\beta_{\ell-1}]$ est la matrice dont la j^e colonne est l'image q-aire du produit $\alpha^i\beta_j \in GF(q^a)$.

On voit aisément que les colonnes de A_i vérifient les propriété 1° et 2° du théorème 1. Pour montrer les conditions 3° et 4°, il suffit d'observer d'abord que les colonnes de A_i (vues comme des triplets d'éléments de $GF(q^a)$) sont contenues dans le sous-espace W_i de $[GF(q)]^{\ell-1} = [GF(q^a)]^3$ formé de tous les triplets d'images q-aires des solutions de l'équation

$$(*) \qquad x_0^2 - x_1 x_2 = 0$$

qui sont parallèles à la solution particulière $[1,\alpha^i,\alpha^{-i}]^T$. Ensuite, le fait que la conique non-dégénérée (*) définisse un (q^a+1)-arc dans le plan projectif $PG(2,q^a)$ sur le corps $GF(q^a)$ entraîne immédiatement que toutes les sommes $W_{i_1}+W_{i_2}+W_{i_3}$ de sous-espaces W_i distincts sont directes. Enfin, il est évident par construction que le poids minimum de W_i est ≥ 3. On peut exprimer ce qui précède dans le résultat suivant.

Théorème 2

Soit q une puissance d'un nombre premier et k un entier naturel quelconque. Si $\ell = 3k+2$, $m \leq q^{2k+1}$ et α est un élément primitif de $GF(q^{2k+1})$, alors les égalités (1), (2) et (3) où $i=0,1,\ldots,m-1$ définissent une matrice de contrôle H d'un code $C(q,\ell,\alpha,m)$ 2-correcteur adapté aux systèmes d'information par sous-blocs de longueur ℓ.
Le taux de ce code, $R = \frac{(\ell-1)m}{(m+2)}$, est très élevé aussitôt que m est suffisamment près de q^{2k+1} (pour $k \geq 4$, $R > 90\%$). De plus C permet de corriger tous les schémas d'erreurs du type i) erreur simple dans 3 sous-blocs d'information ou ii) erreur double dans un sous-bloc d'information et erreur simple dans un autre.

2. - Décodage

Soit y=x+e le mot reçu, e étant le vecteur erreur. On a $y= y_0^T, \ldots, y_{m+1}^T]^T$ où $y_i=[y_{i1}, \ldots, y_{i\ell}]^T$ est le i^e sous-bloc (a^T désigne le transposé de a). Le syndrôme de y est

$$Hy = He = \begin{bmatrix} s_r \\ s_c \\ s_p \end{bmatrix} \quad \text{où}$$

$$\begin{cases} s_r = [\sum_{j=1}^{\ell} e_{ij} \mid i=0,\ldots,m-1]^T & \text{est le syndrôme-rangée,} \\ s_c = \sum_{i=0}^{m-1} A_i e_i + [e_m^T, e_{m+1}^T \mid_{\ell-1}]^T & \text{est le syndrôme-colonne et} \\ s_p = \sum_{i=0}^{m-1} e_{i\ell} + e_{m+1,\ell} & \text{est le syndrôme-parité} \end{cases} \qquad (4)$$

($e_{m+1}|_{\ell-1}$ désigne ici le vecteur formé des $\ell-1$ premières composantes de e_{m+1}).

Etant donnée la forme partitionnée des matrices A_i, on peut écrire

$$s_c = \begin{bmatrix} s_c^{(1)} \\ \\ s_c^{(2)} \\ \\ s_c^{(3)} \end{bmatrix} = \begin{bmatrix} \sum_{i=0}^{m-1} N_0 e_i \\ \\ \sum_{i=0}^{m-1} \alpha^i (N_0 e_i) \\ \\ \sum_{i=0}^{m-1} \alpha^{-i} (N_0 e_i) \end{bmatrix} + \begin{bmatrix} \\ e_m \\ \\ e_{m+1|\ell-1} \\ \\ \end{bmatrix} \qquad (5)$$

On peut alors distinguer plusieurs cas.

<u>Cas 1</u> : $W(s_r) = 0$.

Ce cas survient lorsque

a) il n'y a pas eu d'erreur

b) il y a eu erreur double dans un sous-bloc d'information ou

c) il y a eu erreur simple ou double dans les sous-blocs de parité.

Tous ces cas doivent être distingués par les syndrômes s_c ou s_p.

Dans le cas b), s'il y a eu erreur double dans le i^e sous-bloc d'information et si $s_p=0$, alors en vertu de (4) et (5) $s_{ri}=\xi+\eta= 0$ où $\xi,\eta\in GF(q)$ et

$$s_c^{(1)} = N_0 e_i = \xi N_{0u} + \eta N_{0v}$$

$$s_c^{(2)} = \alpha^i N_0 e_i = \alpha^i (\xi N_{0u} + \eta N_{0v}).$$

On a donc $\alpha^i = s_c^{(2)}[s_c^{(1)}]^{-1}$ dans le corps $GF(q^a)$ ce qui détermine i. Ensuite u,v,ξ et η sont déterminés en localisant $s_c^{(1)}$ dans une table contenant les $(q-1)\ell(\ell-1)$ combinaisons linéaires des colonnes de N_0 dont la somme des coefficients est s_{ri}.

D'un autre côté, si dans le cas b) nous avions $s_p=\eta\neq 0$ alors i serait déterminé comme plus haut et il y aurait eu erreur dans le symbole de parité du i^e sous-bloc d'information et une erreur dans le k^e symbole défini par $s_c^{(1)} = \xi N_{0k}$ du même sous-bloc.

Dans le cas c), le poids de s_c est 1 ou 2. Cela distingue le cas c) du cas b), puisque dans ce dernier cas le poids de s_c est au moins 3 par hypothèse.

Cas 2 : $W(s_r)=1$

Ce cas survient lorsqu'il y a une ou deux erreurs dans le i^e sous-bloc identifié par $s_{ri}=\xi_1+\xi_2\neq 0$ et

a) aucune autre erreur

b) une erreur dans les sous-blocs de parité

c) une erreur double dans un autre sous-bloc d'information.

Dans le cas a), $s_c^{(1)}=\xi_1 N_{0u_1} + \xi_2 N_{0u_2}$ identifie la position en erreur. Dans le cas b), on a

$$\begin{cases} s_c^{(1)} = \xi_1 N_{0u_1} + \xi_2 N_{0u_2} + \eta\varepsilon^{(1)} \\ s_c^{(2)} = \alpha^i(\xi_1 N_{0u_1} + \xi_2 N_{0u_2}) + \eta\varepsilon^{(2)} \\ s_c^{(3)} = \alpha^{-i}(\xi_1 N_{0u_1} + \xi_2 N_{0u_2}) + \eta\varepsilon^{(3)} \end{cases} \quad \text{où} \quad \begin{bmatrix} \\ e_m \\ \\ e_{m+1}|\ell-1 \\ \end{bmatrix} = \begin{bmatrix} \eta\varepsilon^{(1)} \\ \eta\varepsilon^{(2)} \\ \eta\varepsilon^{(3)} \end{bmatrix}$$

est un vecteur de poids 1, les vecteurs $\varepsilon^{(j)}$, de taille $2k+1$, ne contenant que 0 ou 1. Trois sous-cas sont possibles

$$\varepsilon^{(1)} = \varepsilon^{(2)} = 0 \quad \text{i.e} \quad s_c^{(2)} = \alpha^i s_c^{(1)} \quad \text{et} \quad s_c^{(3)} - \alpha^{-i} s_c^{(1)} = \eta\varepsilon^{(3)}$$

$$\varepsilon^{(1)} = \varepsilon^{(3)} = 0 \quad \text{i.e} \quad s_c^{(3)} = \alpha^{-i} s_c^{(1)} \quad \text{et} \quad s_c^{(2)} - \alpha^{i} s_c^{(1)} = \eta\varepsilon^{(2)}$$

$$\varepsilon^{(2)} = \varepsilon^{(3)} = 0 \quad \text{i.e} \quad s_c^{(2)} = \alpha^{2i} s_c^{(3)} \quad \text{et} \quad s_c^{(3)} - \alpha^{-2i} s_c^{(2)} = \eta\varepsilon^{(2)}$$

Dans tous ces sous-cas ξ_1, ξ_2, u_1, u_2 et η sont univoquement déterminés. Dans le dernier cas c), on peut écrire

$$s_c = \xi_1 A_{iu_1} + \xi_2 A_{iu_2} + (\eta A_{jv} + \zeta A_{jt})$$

(erreur double dans le j^e sous-bloc, $v\neq t$). Cette relation détermine j puisque si $j'\neq j$ et si

$$s_c = \xi_1' A_{iu_1'} + \xi_2' A_{iu_2'} + (\eta' A_{j'v'} + \zeta' A_{j't'}) \quad , \quad \text{alors}$$

$\sum_{k=1}^{2} (\xi_k A_{iu_k} - \xi'_k A_{iu'_k}) = (\eta A_{jv} + \zeta A_{jt}) - (\eta' A_{j'v'} + \zeta' A_{j't'}) \in W_i \cap (W_j + W_{j'})$ ce qui entraîne que $\sum_{k=1}^{2} (\xi_k A_{iu_k} - \xi'_k A_{iu'_k}) = 0$ (puisque la somme $W_i + W_j + W_{j'}$, est directe) et finalement $v=t$ une contradiction.

Pour trouver effectivement, j, u_1, u_2, v et t, on résout dans le corps $GF(q^{2k+1})$ le système linéaire suivant pour les 2 variables $(\xi_1 N_{0u_1} + \xi_2 N_{0u_2})$ et $(\eta N_{0v} + \zeta N_{0t})$

$$(*)\begin{cases} s_c^{(1)} = (\xi_1 N_{0u_1} + \xi_2 N_{0u_2}) + (\eta N_{0v} + \zeta N_{0t}) \\ s_c^{(2)} = (\xi_1 N_{0u_1} + \xi_2 N_{0u_2}) + \alpha^j (\eta N_{0v} + \zeta N_{0t}) \\ s_c^{(3)} = (\xi_1 N_{0u_1} + \xi_2 N_{0u_2}) + \alpha^{-j} (\eta N_{0v} + \zeta N_{0t}) . \end{cases}$$

La condition de compatibilité de ce système donne

$$\frac{\alpha^{-i}[s_c^{(2)} - \alpha^i s_c^{(1)}]}{\alpha^i s_c^{(1)} - s_c^{(3)}}$$

et détermine j. Résolvant (*), on trouve $\xi_1 N_{0u_1} + \xi_2 N_{0u_2}$ et $\eta N_{0v} + \zeta N_{0t}$ ce qui permet de retrouver ξ_1, ξ_2, u_1, u_2, η, ζ, v et t puisque les sommes $\xi_1 + \xi_2 = \xi = s_{ri}$ et $\eta + \xi = 0 = s_{rj}$ sont connues.

Cas 3 : $W(s_r) = 2$.

Les deux coordonnées non-nulles $s_{ri} = \zeta = \zeta_1 + \zeta_2$ et $s_{rj} = \eta = \eta_1 + \eta_2$ de s_r détermine les sous-blocs en erreur. Pour trouver les positions en erreur, on résout dans le corps $GF(q^{2k+1})$ le système

$$s_c^{(1)} = (\zeta_1 N_{0u_1} + \zeta_2 N_{0u_2}) + (\eta_1 N_{0v_1} + \eta_2 N_{0v_2})$$

$$s_c^{(2)} = \alpha^i (\zeta_1 N_{0u_1} + \zeta_2 N_{0u_2}) + \alpha^j (\eta_1 N_{0v_1} + \eta N_{0v_2})$$

$$s_c^{(3)} = \alpha^{-i} (\zeta_1 N_{0u_1} + \zeta_2 N_{0u_2}) + \alpha^{-j} (\eta_1 N_{0v_1} + \eta_2 N_{0v_2})$$

après avoir vérifié la condition de compatibilité

$$\alpha^i \alpha^j s_c^{(3)} + s_c^{(2)} - (\alpha^i + \alpha^j) s_c^{(1)} = 0 .$$

on trouve alors les vecteurs $\zeta_1 N_{0u_1} + \zeta_2 N_{0u_2}$ et $\eta_1 N_{0v_1} + \eta_2 N_{0v_2}$ ce qui détermine ζ_1, ζ_2, u_1, u_2, η_1, η_2, v_1 et v_2 puisque les sommes $\zeta_1+\zeta_2=\zeta$ et $\eta_1+\eta_2=\eta$ sont connues.

Cas 4 : $W(s_r) = 3$

Les trois coordonnées non-nulles $s_{ri}=\xi$, $s_{rj}=\eta$, $s_{rj'}=\zeta$ indiquent les 3 sous-blocs en erreur. Pour déterminer les positions en erreur on résout dans $GF(q^{2k+1})$ le système inversible

$$\begin{cases} s_c^{(1)} = \xi N_{0u} + \eta N_{0v} + \zeta N_{0t} \\ s_c^{(2)} = \alpha^i \xi N_{0u} + \alpha^j \eta N_{0v} + \alpha^{j'} \zeta N_{0t} \\ s_c^{(3)} = \alpha^{-i} \xi N_{0u} + \alpha^{-j} \eta N_{0v} + \alpha^{-j'} \zeta N_{0t}. \end{cases}$$

Remarque :

La méthode naïve présentée ici pour décoder la classe de codes présentée exige de garder en mémoire q-1 tables contenant $(q-1)\ell(\ell-1)$ entrées, chaque entrée comportant 3 vecteurs de $[GF(q)]^\ell$ et un élément de GF(q). Les calculs de syndromes, de l'ordre de $m \le q^a$ avec $a = (2\ell-1)/3$, sont effectués dans le corps GF(q) au fur et à mesure de la réception des sous-blocs. Dans le pire des cas, le décodage exige la résolution d'un système linéaire de 3 équations à 3 inconnues à coefficients dans le corps $GF(q^a)$.

Voici quelques chiffres dans le cas binaire

longueur ℓ des sous-blocs	5	8	11	14	17
table à garder en mémoire en octets	60	168	480	910	1900
arithmétique dans le corps à q^a éléments	2^3	2^5	2^7	2^9	2^{11}

Remerciements :

Nous remercions Messieurs Gérard Séguin et Vijay Bhargava de nous avoir donné accès au document [2] qui a motivé notre étude.

Bibliographie

[1] G. Séguin, P.E. Allard, V.K. Bhargava, "A class of High Rate Codes for Byte-Oriented Information Systems", IEEE Trans. on Communication Theory, vol. com-31, n° 3, March 1983.

[2] P.E. Allard, V.K. Bhargava, G. Séguin, "Réalization, Economic and Performance Analysis of Error-Correcting codes and ARQ Systems for Broadcast Telidon and other Videotex Systems", Dept. Elec. Eng., Concordia University, Montréal, P.Q. Canada, Res. Dept. June 1981.

[3] L.A. Dunning, M.R. Varanasi, "Code Constructions for Error Control in Byte Organized Memory Systems", IEEE Trans. on Computers, vol. c-32, n° 6, June 1983.

[4] J. Ziv, A. Lempel, "A Universal Algorithm for Sequential Data Compression", IEEE Trans. Information Theory, vol. IT-23, n° 3, May 1977, 337-343.

[5] T.A. Welch, "A Technique for High-Performance Data Compression" , computer, June 1984, 8-19.

LOUSTICC SIMULATION SOFTWARE : EXPERIMENTAL RESULTS OF CODING SYSTEMS

M.C. Gennero

AAECC Lab.
Université Paul Sabatier
118 route de Narbonne
31062 Toulouse cédex/France

SUMMARY

We present simulation results from our software program LOUSTICC. It can simulate data transmissions through a noisy channel.

The given results present an evaluation of the execution time for a data transmission simulation.

INTRODUCTION

One of the problems of coding system designing in data communication is to test the efficiency of the choosen error-correcting codes. G.C. Clark and J.B. Cain {6}, and E.R. Berlekamp also {1} show that it is necessary to do preliminary simulations. In fact, for most used channels few (or no) indications about error kind and repartition are available.

To this aim (testing coding efficiency), we have conceived and realized a software program : LOUSTICC. It includes several coding modules, a module to add transmission errors whose kind and repartition are modelized. It includes also several decoding modules according to the coding ones.

As far as we know, there are few publications about such softwares. Let us mention for example :

- A software (conceived in CII-HB {5}) to protect data on magnetic disks. It has only one coding module for Reed-Solomon codes, and no error modelization.
- The ANIS simulator {12} realized in the NADIR pilot project (INRIA Lab). Anis was used to simulate transmissions using

European telecommunication satellite TELECOM1. This simulator is a complete system including retransmission protocols but no error modelization. Only the bit error rate is a parameter for Anis.

Other papers give simulation results, but the program is not the purpose of the paper {13}{16}.

In the following we don't present LOUSTICC, because it has already been published {8}{9}{10} , but we give simulation principles and simulation results.
These results are obtained from the basic version of LOUSTICC, and from another version specific to Fire codes. This second version makes it possible to get a very fast simulation of binary frames of length up to 2.5 Mbits.
Another version of LOUSTICC exists to simulate code concatenation and interleaving for Reed-Solomon and convolutional codes.

THE SIMULATION PROGRAM

The software program (LOUSTICC) simulates data transmission through a noisy channel. It performs coding (by error-correcting codes), addition error patterns, and decoding.

1) Data protection

Data protection is ensured by error-correcting codes. The coding here is systematic and uses block codes. For the moment the interleaving and concatenation techniques are not simulated in LOUSTICC.
As a matter of fact block codes are used, each one having its own decoding scheme :

- *Fire codes* (shortened or not) {17}. For these codes we use the Meggitt decoding technique {2}.
- *Reed-Solomon codes* over fields $\mathbb{F}_p$ (p prime). The decoding scheme uses the algorithm proposed by Reed, Miller et al. in {20}.
- *The extended binary Golay (24,12,8) code*. Its decoding algorithm uses the permutation group of the code {11}{21}. This code corrects three independent errors.
- *One self-dual (64,32,12) code*. In this case we use a thre-

shold decoding proposed by G.L. Feng {7}, with a variant. With this decoding algorithm, the code corrects only two errors, but it is powerful for error detection.

2) Addition of noises

To simulate transmission errors, we add error patterns to the coded frame (consecutive codewords).

In LOUSTICC we have modilized the average number of binary symbols free of error, between two consecutive error pattern beginnings. The errors can be independent or occur in bursts. Their intervals are modelized by the random variable :

$$Y = a\,X^{i} + t$$

Where X is a random variable of three possible kinds : an uniform, or a geometric, or a normal one (approximation of a normal by a binomial one).
The a, i and t parameters are not randomly choosen, but they are computed to adjust the experimental error rate to a choose "theoretical" one.
Th Y random variable give the number of sysbols without errors (bursts or no) between two errors (beginning of the burst or position of independent error).
The theoretical binary error rate is given by :

$\tau_t = 1 \;/\; (1 + a\,E(X^{i}) + t)$ for independent errors

$\tau_t = \ell \;/\; 2(1 + a\,E(X^{i}) + t)$ for bursts, where ℓ is the maximal burst length

3) LOUSTICC output

After a LOUSTICC run is completed, we obtain :

- *The binary experimental error rate.* It is equal to the ratio of the erroneous binary symbol number -to- the transmitted binary symbol number.
- *The binary residual error rate.* It is equal to the ratio of the erroneous binary symbol number (after decoding) -to- the total binary symbol number.
- *The transmission efficiency.* It is equal to the ratio of binary symbol number of the file (to be transmitted) -to- the

total binary symbol number really transmitted. This computation includes the fact that the simulated channel is full duplex or simplex (for FD, retransmissions are possible).

- *Statistical results about error repartition* : mean and variance of the random variable X.
- *Two files* used to have a subjective idea of the protection. These two files are on one hand the initial file without coding and with error patterns, and on the other hand the file after coding, error addition and decoding.

Only the two first results are needed to compute the coding gain (if the noises are white Gaussian).

RESULT OF SIMULATION

Now we give simulation results. They give an experimental evaluation of the process time required by LOUSTICC, and the residual error rate versus the bit error rate.

For a LOUSTICC run, the user enters the following different parameters :

- The chosen error-correcting code (and then the decoding algorithm).
- The theoretical bit error rate and the distribution of the random variable X (the program automatically computes the parameters a, i, and t).
- The frame length.

1) Fire codes

The Fire codes are used for burst correction, it is the case for example of magnetic recordings (disks, tapes, memories,...).

a) Recall

The generator polynomial of Fire codes is of the form :

$$g(X) = P_m(X)\ (X^c - 1)$$

where $P_m(X)$ is an irreducible polynomial of degree m, such that $P_m(X)$ don't divide $X^c - 1$.

The length of the code is n = LCM(e,c) (e is the order of $P_m(X)$ roots).

The dimension of the code is k = n - (c+m).

One can use Fire codes in error detection only, or in error correction only, or in both correction and detection. Used in error correction and detection (our case), these codes are able to correct bursts of length L_c and to detect bursts of length L_d ($L_d > L_c$) if :

$$L_c + L_d \leq c + 1$$

$$L_d < m$$

b) results (for Fire codes)

In the table 1 we give the C.P. time in terms of the bloc length.

In the table 2 we give the C.P. time in terms of the input bit error rate.

We use the following notations :

- n and k : the parameters of the code (length and dimension).
- g(x) : the generator polynomial, in octal. As an example $1 + x + x + x + x$ is written 6232.
- L_C, L_D : the detectable (correctable) burst length.
- P_C, P_r : the effective input (output) bit error rate.
- τ : the theoretical input binary error rate.
- C.P. : the process execution time.

n	k	g(X)	L_c	L_d	P_c (10^{-4})	P_r (10^{-4})	C.P. (seconds)
585	560	461563066	7	7	5.16	2.43	42.854
651	620	53020005302	9	13	6.23	2.31	43.235
819	786	4414440044144	11	11	5.99	1.61	43.850
990	956	53020005302	11	12	6.10	2.22	44.484
1105	1084	73503564	7	7	4.80	1.30	40.917
1209	1186	53022541	7	7	4.61	1.36	43.824
1302	1278	530212604	7	8	4.60	1.19	43.499
1365	1332	755140075514	11	11	5.90	1.85	44.781
1430	1396	516240024712	11	12	6.00	1.75	45.339
1533	1503	42040004204	8	14	4.15	0.78	45.236
1638	1612	441451031	7	8	4.61	1.22	45.428
1785	1756	5610000561	7	16	4.71	1.01	47.067
1780	1840	73500003564	7	16	4.55	1.48	48.490
1958	1925	757100036744	10	13	6.50	1.99	47.555

table 1 : C.P. time as a function of the bloc length for Fire codes.

Remark : The used Fire codes (tables 1 and 2) have a transmission rate larger than 95 %. They are determined by another software realized in the AAECC lab.

τ	P_c (10^{-4})	P_r (10^{-4})	C.P. (seconds)
12.00	13.20	4.14	47.189
20.00	21.40	6.84	57.994
28.50	30.20	12.50	64.031
40.00	40.80	2.19	71.296
50.00	52.10	34.80	78.723
1.00	1.49	0.35	40.364
1.10	1.29	0.35	41.082
1.20	1.25	0.49	41.062
1.30	1.44	0.33	41.089
2.00	2.70	0.70	41.915
2.99	3.45	1.15	42.413
4.07	4.98	1.58	43.872
5.00	5.08	2.10	44.072
8.00	8.20	3.27	47.520
0.90	1.04	0.51	41.066
0.95	0.96	0.26	40.920

table 2 : C.P. time as a function of the input bit error rate for Fire code (910,884) $L_c=7$, $L_d=8$, $g(X)=755157323$.

2) Reed-Solomon codes

a) Recall

The Reed-Solomon codes are also used to correct and detect bursts.

The Reed-Solomon codes are defined over $GF(p^s)$, p a prime. Their generator polynomial is defined by its roots :

$$g(X) = \prod_{i=0}^{d-2} (X - \alpha^{m+i}), \text{ where } \alpha^m \text{ is a primitive element}$$

of $GF(p^s)$.

The length of a Reed-Solomon code is $n = p^s - 1$,

The minimal distance is d

Its dimension is $k = n - d + 1$

Usually one uses Reed-Solomon codes over $GF(2^s)$. Each symbol of the field is encoded with s binary symbols.

One can also use Reed-Solomon codes over GF(p), p a prime. In the LOUSTICC program we use Reed-Solomon codes over GF(p), with p equal to 2^s-1. Each group of s binary symbols forms a Reed-Solomon symbol. The value

2^m-1 is encoded by zero. At the decoding this value (zero) is considered as an error and can then be corrected.

b) Results for Reed-Solomon codes.

The notations are the same as previously.

P_c (10^{-3})	10.660	9.695	7.927	7.660	7.565	6.431	4.994	4.818	4.701
P_r (10^{-3})	9.450	7.922	4.454	4.915	4.238	2.617	1.539	6.460	1.159

Table 3 : binary residual error rate as a function of the binary input error rate, for a Reed-Solomon code (126,120,7).

CONCLUSION

In this paper we have presented some results obtained from an error-correcting codes simulation software. We also gave results about the execution time of runs, with Fire and Reed-Solomon codes. The given results are obtained with correction only (i.e. detection indication is not taken into account). But LOUSTICC can also simulate retransmission techniques in the case of uncorrectable errors are detected (full duplex simulations).

Two particular versions of the software are designed for pratical applications :

- The first one makes it possible to quickly simulate transmissions of very long binary frames (up to 2 Mbits), with Fire codes. It solves a picture transmission simulation problem asked by the MATRA ingineers {22}.
- The second one uses Reed-Solomon codes over $GF(2^s)$ with possible interleaving, concatenated with convolutionnal codes. It is in realization for a contract wiht CNES. This software simulates additive white Gaussian noise. In this case we are able to plot the curves giving the error probability after decoding versus E_b/N_o ratio {18}{19}.

REFERENCES

1 BERLEKAMP E.R.
"The technology of error correcting codes"
IEEE Trans. Inf. Theory, vol 68 n°5, May 1980

2 BERLEKAMP E.R.
"Algebraic coding theory"
Mac Graw Hill, New York 1968

3 BEST M.R., ROEFS H.F.A.
"Concatened coding on spacecraft-to-ground telemetry channe : méthod"
IEEE Conference on Comm. June 1981

4 BETH T., SAGERER G.
"CODEC : A program system for interactive development of error correcting coders/decoders"
Journal of Information processing and cybernetics EIK 17, 1981 pp 145-147

5 BOULENOUARD A.
"Application des codes de Reed-Solomon à la correction d'erreurs dans les unités à disques magnétiques"
Thèse de Docteur de 3° Cycle (Université Paris Sud), Juillet 1982

6 CLARK G.C., CAIN J.B.
"Error-correction coding for digital communications"
Plenum Press, New York (ISBN 0-306-40615-2 1981

7 FENG G.L.
"Generalized threshold decoding of cyclic codes"
Acts of AAECC-1, Discrete Math vol 56 n°2 & 3, 1985 pp 147-154

8 GENNERO M.C., POLI A., THIONG LY J.A.
"Codes correcteurs en transmission très bruitée : résultats de simulations expérimentales"
9ième Colloque International GRETSI, Nice 1983

9 GENNERO M.C.
"Un logiciel de simulation de transmission d'information : LOUSTICC"
Mémoire d'Ingénieur CNAM (Informatique), Centre de TOULOUSE (1983)

10 GENNERO M.C., PAPINI O.
"Utilization of error-correcting codes for data transmission simulations"
Acts of AAECC-1, Discrete Math. Vol 56 n°1 & 2, 1985 pp155-168

11 GORDON D.L.
"Minimal permutations sets for decoding the binary Golay code"
IEEE Trans. Inform. Theory vol IT-28, pp 541-543, 1982

12 GRANGE J.L., HUITEMA C., ZIMMERMANN H.
"Utilisation informatique des satellites de télécommunication, identification des problèmes posés, et éléments de solution"
Projet Pilote NADIR (INRIA), Ref GEN 3.500, Décembre 1980

13 HELLER J.A., JACOB I.M.
"Viterbi decoding for satellite and space communication"
IEEE Trans. Comm. Technology vol COM-19 n°15, 1971

14 MACCHI C., GUILBERT J.F.
"Téléinformatique, transport et traitement de l'information dans les réseaux et systèmes téléinformatiques"
Editions DUNOD Informatique, 2ième édition Avril 1983

15 MACWILLIAMS F.J., SLOANE N.J.A.
"The theory of error-correcting codes"
North Holland Publishing Company 1977

16 MODESTINO, MATIS
"Interactive simulation of digital communications"
IEEE Selected areas on Comm. 1984 January

17 PETERSON W.W.
"Error correcting codes"
MIT Press, Cambridge, Mass 1961

18 POLI A., RANDRIANANJA D., THIONG LY J.A.
"Codage Reed-Solomon entrelacé, concaténé à un codage convolutionnel"
Rapport de contrat intermédiare CNES (Janvier 1984)

19 POLI A., RIGONI C., RANDRIANANJA D.
"Décodeur du code convolutionnel"
Rapport de contrat intermédiare CNES (Mai 1985)

20 REED I.S., TRUONG T.K., MILLER R.L.
"The fast decoding of Reed-Solomon codes using Fermat theoretic transforms and continued fractions"
IEEE Trans. Inform. Theory vol 24 n°1, 1978

21 WOLFMANN J.
"A permutation decoding of the (24,12,8) Golay code"
IEEE Trans. Inform. vol IT-29 n°5, Sept 1983, pp 748-750

22 MATRA Espace (private correspondence)

AN ALGORITHM OF COMPLETE DECODING OF DOUBLE-ERROR-CORRECTING GOPPA CODES

G.L.Feng
Shanghai Institute of Computer Technology, P.R.China
K.K.Tzeng
Department of Computer Science and Electrical Engineering,
Lehigh University, Pennsylvania 18015, U.S.A.

I. Introduction

As perfect codes are known to be scarce, there has been much effort directed to the investigation of nearly-perfect codes and quasi-perfect codes [1]. Early in 1960, Gorenstein, Peterson and Zierler [2] showed that all double-error-correcting primitive binary BCH codes are quasi-perfect. In 1971, Hartmann [3] showed a method of complete decoding of such codes. In 1981, Moreno [4] proved that, when m is odd, the binary Goppa codes with parameters $(2^m, 2^m-2m, 5)$ are also quasi-perfect. More recently, Feng and Tzeng [5] proved that, for any syndrome, except the case where m is even and the syndrome terms are $s_1=0$ and $s_3=1$, the corresponding coset is of weight ≤ 3. When m is even, the coset corresponding to $s_1=0$ and $s_3=1$ is shown to be of weight 4. Therefore, in addition to proving that the double-error-correcting Goppa codes with odd values of m are quasi-perfect, that the codes with even values of m are nearly-quasi-perfect was also proved. In the paper [5], a complete decoding procedure of such Goppa codes was also shown. In this paper, we have shown an algorithm of the complete decoding procedure. The algorithm of complete decoding requires at most m times calculating inner product of vectors over GF(2) and finding roots of quadratic equation in $GF(2^m)$.

II. Complete Decoding Procedure

In this paper, we consider the binary Goppa codes with parameters $(2^m, 2^m-2m, 5)$ and generator polynomial $G(z)=z^2+z+\beta$. Complete decoding of such Goppa codes is that, for any syndrome terms s_1 and s_3, to find the minimum positive integer t and values $x_1, x_2, \cdots, x_k$ in $GF(2^m)$ such that

$$\sum_{i=1}^{t} \frac{1}{x_i^2+x_i+\beta} = s_1$$
$$\sum_{i=1}^{t} \frac{x_i}{x_i^2+x_i+\beta} = s_3 \qquad (1)$$

It is known from [5] that, if m is odd $t \leq 3$; if m is even and $s_1 \neq 0$ or $s_3 \neq 1$, $t \leq 3$; if m is even and $s_1=0$ and $s_3=1$, $t=4$. A complete decoding procedure of such Goppa codes can be formulated as follows:

(1) If $s_1=0$ and $s_3=0$, then there are no errors.

(2) If $s_1 \neq 0$ and $(s_3/s_1)^2+(s_3/s_1)+\beta = 1/s$, then t=1 and $x_1=s_3/s_1$.

(3a) If $s_1 \neq 0$, $s_3=0$ and $tr(1/\beta s_1^2)=1$, then two errors occurred. First find x_2 from Eq.(2), then x_1 from Eq.(3).

$$(s_1+\beta s_1^2)x_2^2+\beta s_1^2 x_2+\beta(s_1+\beta s_1^2)=0 \qquad (2)$$

$$x_1=(s_1 y_2+x_2)/(s_1 y_2+1) \qquad (3)$$

where $y_i=x_i^2+x_i+\beta$ for i=1,2.

(3b) If $s_1=0$, $s_3 \neq 0$ and $tr(1/s_3)=1$, then two errors occurred. First find x_2 from Eq.(4), then x_1 from Eq.(3):

$$s_3^2 x_2^2+s_3^2 x_2+(s_3^2\beta+s_3)=0 \qquad (4)$$

(4a) If $s_1 \neq 0, 1/\beta$, $s_3=0$ and $tr(1/\beta s_1^2)=0$, then three errors occurred. first determine θ_0, θ_1 from Eq.(5), then find x_1 from Eq.(6), then x_2 from Eq.(7) and finally x_3 from Eq.(8):

$$\frac{\beta^{\frac{1}{2}}+1/s_1^{\frac{1}{2}}+1/s_1}{\theta_0}+\frac{1/\beta^2 s_1^3}{\theta_1}=1+\frac{1}{\beta s_1} \qquad (5)$$

where $tr(\theta_0)=0$, $tr(\theta_1)=1$.

$$x_1^2+\left(1+\frac{1}{\beta s_1}+\frac{1}{\beta^2 s_1^3\theta_1}\right)x_1+\left(\beta+\frac{1}{\beta s_1^2}+\frac{1}{\beta^2 s_1^4\theta_1}\right)=0 \qquad (6)$$

$$(p+y_1 s_1)x_2^2+(1+p)x_2+p(\beta+1/s_1)=0 \qquad (7)$$

where $p=y_1\beta s_1^2+s_1 x_1$

$$x_3=\left(s_3+\frac{x_1}{y_1}+\frac{x_2}{y_2}\right)/\left(s_1+\frac{1}{y_1}+\frac{1}{y_2}\right) \qquad (8)$$

(4b) If $s_1=0$, $s_3 \neq 0,1$ and $tr(1/s_3)=0$, then three errors occurred. First determine θ_1, θ_2 from Eq.(9), then find x_1 from Eq.(10), then x_2 from Eq.(11) and finally x_3 from Eq.(8)

$$\frac{1+s_3}{s_3^4}=\theta_1\cdot\theta_2 \qquad (9)$$

where $tr(\theta_1)=tr(\theta_2)=1$.

$$x_1^2+x_1+\left(\beta+\frac{1}{s_3}+\frac{1}{s_3^2}+\frac{1+s_3}{s_3^4\theta_1}\right)=0 \qquad (10)$$

$$x^2+\left(1+\frac{1}{q s_3^2}\right)x_2+\left(\beta+\frac{1}{s_3}+\frac{1+x_1}{q s_3^2}\right)=0 \qquad (11)$$

where $q=y_1+1/s_3$.

(5) If $s_1=0$, $s_3=1$ and m is even, then 4 errors occurred. First determine u,v such that $1/u+1/v=1$ and $tr(u)=tr(v)=1$. Then find x_1, x_2 from Eq.(12) and x_3, x_4 from Eq.(13):

$$x_1^2+x_1+\beta = x_2^2+x_2+\beta = u \qquad (12)$$

$$x_3^2+x_3+\beta = x_4^2+x_4+\beta = v \qquad (13)$$

In the above complete decoding procedure, except Eqs.(5) and (9), all equations are quadratic equations over $GF(2^m)$. A formula of roots of a quadratic equation was shown in [6]. Certainly, we can also find roots of a quadratic equation by m linear equations over GF(2). Therefore the key problem of the above complete decoding procedure is how to find roots of Eqs.(5) and (9). In the section III, we shall show an algorithm of finding roots of Eqs.(5) and (9). The algorithm requires at most m times calculating inner product of vectors over GF(2).

III. Solving Eq.(5) and Eq.(9)

Let $1, \beta, \cdots, \beta^{m-1}$ be a basis of $GF(2^m)$. Every element x in $GF(2^m)$ can be expressed by $x_0 1 + x_1\beta + \cdots + x_{m-1}\beta^{m-1}$, where $x_j \in GF(2)$. In the following, we shall denote $(x_0, x_1, \cdots, x_{m-1})$ as X. Let $A=(a_0, a_1, \cdots, a_{m-1})$, we have

$$tr(A\,X)=\sum_{j=0}^{m-1}\sum_{i=0}^{m-1} f_j^{(i)} x_i a_j \quad , \quad \text{for } f_j^{(i)} \in GF(2) \tag{14}$$

Let

$$F=\begin{bmatrix} f_0^{(0)} & f_0^{(1)} & \cdots & f_0^{(m-1)} \\ f_1^{(0)} & f_1^{(1)} & \cdots & f_1^{(m-1)} \\ \vdots & \vdots & & \vdots \\ f_{m-1}^{(0)} & f_{m-1}^{(1)} & \cdots & f_{m-1}^{(m-1)} \end{bmatrix}$$

Lemma 1: Matrix F is nonsigular, namely det $F \neq 0$.

Proof: If det F=0, there must be $c_0, c_1, \cdots, c_{m-1}$ from GF(2), not all zero,such that

$$F\cdot\begin{bmatrix} c_0 \\ c_1 \\ \vdots \\ c_{m-1} \end{bmatrix} = \begin{bmatrix} 0 \\ 0 \\ \vdots \\ 0 \end{bmatrix}$$

Let $C=(c_0, c_1, \cdots, c_{m-1}) \in GF(2^m)$, we have

$$\begin{bmatrix} tr(C\cdot 1) \\ tr(C\cdot\beta) \\ \vdots \\ tr(C\cdot\beta^{m-1}) \end{bmatrix} = \begin{bmatrix} 1 & 0 & \cdots & 0 \\ 0 & 1 & \cdots & 0 \\ \vdots & & & \vdots \\ 0 & 0 & & 1 \end{bmatrix} \cdot F \cdot \begin{bmatrix} c_0 \\ c_1 \\ \\ c_{m-1} \end{bmatrix} = \begin{bmatrix} 0 \\ 0 \\ \vdots \\ 0 \end{bmatrix}$$

Therefore for any element $A \in GF(2^m)$, $tr(C\cdot A)=tr(\sum_{j=0}^{m-1} a_j C\cdot\beta^j)=\sum_{j=0}^{m-1} a_j tr(C\cdot\beta^j)$ =0. On the other hand, $C\neq 0$. It is impossible. So det $F\neq 0$. Q.E.D.

Lemma 2: There must be m elements $Y_1, Y_2, \cdots, Y_m$ from $GF(2^m)$, where $tr(Y_j)=1$, such that det $Y^*\neq 0$, where $Y_i^{-1}=(y_{0i}, y_{1i}, \cdots, y_{m-1.i})$ and

$$Y^*=\begin{bmatrix} y_{01} & y_{11} & \cdots & y_{m-1\cdot 1} \\ y_{02} & y_{12} & \cdots & y_{m-1\cdot 2} \\ \vdots & \vdots & & \vdots \\ y_{0m} & y_{1m} & & y_{m-1.m} \end{bmatrix}$$

Proof: Let $Y_1, Y_2, \cdots, Y_p$ be all elements with trace 1 of $GF(2^m)$, where p is the number of these elements. Let $Y_j^{-1}=(y_{0j}, y_{1j}, \cdots, y_{m-1.j})$ and

$$P=\begin{bmatrix} Y_{01} & Y_{11} & \cdots & Y_{m-1,1} \\ Y_{02} & Y_{12} & \cdots & Y_{m-1,2} \\ \vdots & \vdots & & \vdots \\ Y_{0p} & Y_{1p} & \cdots & Y_{m-1,p} \end{bmatrix}$$

If there are not m elements such that Lemma 2 holds, then there must be $c_0^*, c_1^*, \cdots, c_{m-1}^*$ from GF(2), not all zero, such that

$$P \begin{bmatrix} c_0^* \\ c_1^* \\ \vdots \\ c_{m-1}^* \end{bmatrix} = \begin{bmatrix} 0 \\ 0 \\ \\ 0 \end{bmatrix} \tag{15}$$

From Lemma 1, we know that there must be $c_0, c_1, \cdots, c_{m-1}$ from GF(2) such that

$$F \begin{bmatrix} c_0 \\ c_1 \\ \vdots \\ c_{m-1} \end{bmatrix} = \begin{bmatrix} c_0^* \\ c_1^* \\ \\ c_{m-1}^* \end{bmatrix} \tag{16}$$

From Eq.(15) and Eq.(16), we have $tr(C \cdot Y_j^{-1})=0$ for $j=1,2,\cdots,p$, namely $C=Y_j Z_j$, where $tr(Z_j)=0$, for $j=1,2,\cdots,p$. Therefore non-zero element C can not be expressed by $\theta_1 \cdot \theta_2$, where $tr(\theta_1)=tr(\theta_2)=1$. It is in contradic-tion with Lemma 3 [5] . Q.E.D.

Theorem 1: For any nonzero element A of $GF(2^m)$, there is at least an element Y_j $(1 \le j \le m)$, such that $tr(A \cdot Y_j^{-1})=1$.

Proof: Since

$$\begin{bmatrix} tr(A \cdot Y_1^{-1}) \\ tr(A \cdot Y_2^{-1}) \\ \vdots \\ tr(A \cdot Y_m^{-1}) \end{bmatrix} = Y^* F \cdot \begin{bmatrix} a_0 \\ a_1 \\ \vdots \\ a_{m-1} \end{bmatrix}$$

and $A \neq 0$, namely $a_0, a_1, \cdots, a_{m-1}$ are not all zero. And $Y*F$ is a nonsingular matrix from Lemma 1 and Lemma 2. So there is at least an element Y_j, where $1 \le j \le m$, such that $tr(A\ Y_j^{-1})=1$. Q.E.D.

Let $Y^{**}=Y*F$ and $(Y_{0j}^*, Y_{1j}^*, \cdots, Y_{m-1,j}^*)$ be the j-th row of Y^{**}. In order to find a root of Eq (9), we have:

Algorithm 1: For $j=1,2,\cdots,m$, check whether the inner product of $(a_0, a_1, \cdots, a_{m-1})$ and $(Y_{0j}^*, Y_{1j}^*, \cdots, Y_{m-1,j}^*)$ is 1. If $(a_0, a_1, \cdots, a_{m-1})\ (Y_{0j}^*, Y_{1j}^*, \cdots, Y_{m-1,j}^*)=1$, then Algorithm 1 halts and Y_j is a root of Eq (9).

Let $A^*=(a_0^*, a_1^*, \cdots, a_{m-1}^*)$ satisfy

$$\begin{pmatrix}1\\1\\\vdots\\1\end{pmatrix}=Y^{**}\begin{pmatrix}a_0^*\\a_1^*\\\vdots\\a_{m-1}^*\end{pmatrix} \tag{17}$$

Lemma 3: If $A\neq A^*$, there is at least $Y_{\cdot j}$ $(1\leq j\leq m)$, such that $tr(A\cdot Y_{\cdot j}^{-1})=0$.

Proof: If for every j $(1\leq j\leq m)$, $tr(A\, Y_{\cdot j}^{-1})=1$, then

$$\begin{pmatrix}1\\1\\\vdots\\1\end{pmatrix}=Y^{**}\begin{pmatrix}a_0\\a_1\\\vdots\\a_{m-1}\end{pmatrix} \tag{18}$$

Comparing Eq (17) and Eq (18), we have $A=A^*$. It is in contradiction with $A\neq A^*$.

For every m, only two cases are possible:

(1) There are $\tilde{\theta}_1$ and $\tilde{\theta}_0$, where $tr(\tilde{\theta}_1)=1$ and $tr(\tilde{\theta}_0)=0$, such that $\tilde{\theta}_1\cdot\tilde{\theta}_0=A^*$.

(2) There are not θ_1 and θ_0, where $tr(\theta_1)=1$ and $tr(\theta_0)=0$, such that $\theta_1\cdot\theta_0=A^*$.

For the equation $A=\theta_1\cdot\theta_0$, where $tr(\theta_1)=1$ and $tr(\theta_0)=0$. (19)
we have:

Algorithm 2: When $A\neq A^*$, for $j=1,2,\cdots,m$ checking whether the inner product of $(a_0,a_1,\cdots,a_{m-1})$ and $(Y^*_{0j},Y^*_{1j},\cdots,Y^*_{m-1,j})$ is zero, we can find a root of Eq (19).

When $A=A^*$, if m satisfies the condition of Case (1), then $\tilde{\theta}_0,\tilde{\theta}_1$ is a root of Eq (19), otherwise Eq (19) has not any root.

Now we consider Eq (5), it is equivalent to

$$\frac{\dfrac{\beta^{\frac{1}{2}}S_1+S_1^{\frac{1}{2}}+1}{S_1}\cdot\dfrac{\beta S_1}{1+\beta S_1}}{\theta_0}+\frac{\dfrac{1}{\beta^2 S_1^3}\cdot\dfrac{\beta S_1}{1+\beta S_1}}{\theta_1}=1 \tag{20}$$

Let $A'=\dfrac{\beta^{\frac{1}{2}}S_1+S_1^{\frac{1}{2}}+1}{S_1}\cdot\dfrac{\beta S_1}{1+\beta S_1}=\beta^{\frac{1}{2}}+\dfrac{\beta}{1+\beta S_1}+\dfrac{\beta^{\frac{1}{2}}}{1+\beta^{\frac{1}{2}}S_1^{\frac{1}{2}}}$ and $B'=\dfrac{1}{\beta S_1^2\,(1+\beta S_1)}$

then $$tr(A')=tr(\beta^{\frac{1}{2}})=1 \tag{21}$$

Eq (20) becomes

$$\frac{A'}{\theta_0}+\frac{B'}{\theta_1}=1$$

namely
$$\theta_1=B'+\frac{A'\,B'}{A'+\theta_0} \tag{23}$$

And let $A'B'=A$. Since $tr(\theta_1)=1$, Eq (23) becomes $tr(B'+\frac{A}{\theta_1^*})=1$, where $\theta_1^*=A'+\theta_0$. If $tr(B')=0$, then $tr(\frac{A}{\theta_1})=1$. We can use Algorithm 1 to find

θ_1^*, so θ_0 and θ_1 are also found.

If $tr(B')=1$, then $tr(\frac{A}{\theta_1^*})=0$. Since Eq.(5) must have root, we can find a root θ_1^* by Algorithm 2.

From the above discussion it is known that at most calculating m times inner product of vectors over GF(2), we can find a root of Eq.(5) or Eq.(9).

IV. Examples

Example 1: For $GF(2^5)$, we can choose $Y_1=\alpha^{11}$, $Y_2=\alpha^{22}$, $Y_3=\alpha^{13}$, $Y_4=\alpha^{26}$ and $Y_5=\alpha^{21}$, where α is a primitive element of $GF(2^5)$ and $\alpha^5+\alpha+1=0$. According to Algorithm 1, for any nonzero element A of $GF(2^5)$, there must be a Y_j $(1\leq j\leq 5)$, which is a root of Eq.(9).

For example, $\theta_1\cdot\theta_2=\alpha^{14}$, using Algorithm 1, we know $\theta_1=\alpha^{11}$ and $\theta_2=\alpha^3$ is a solution.

Example 2: For $GF(2^4)$, we can choose $Y_1=\alpha^7$, $Y_2=\alpha^{14}$, $Y_3=\alpha^{13}$ and $Y_4=\alpha^{11}$, where α is a primitive element of $GF(2^4)$ and $\alpha^4+\alpha+1=0$. So $A^*=1$. According to Algorithm 2, for any A of $GF(2^4)$, where $A\neq 1,0$, then there must be a Y_j $(1\leq j\leq 4)$, which is a root of Eq (19). For A=1, we have $\tilde{\theta}_1=\alpha^{11}$, $\tilde{\theta}_0=\alpha^4$ is a root of Eq.(19).

For example, $\frac{\alpha^3}{\theta_0}+\frac{\alpha^6}{\theta_1}=\alpha$, namely $\frac{\alpha^6}{\theta_0}+\frac{\alpha^9}{\theta_1}=1$. In this case $A=\alpha^6\cdot\alpha^9=1$. So $\theta_1^*=\alpha^{11}$, and using Eq.(11), we have $\theta_0=\alpha^{11}+\alpha^6=\alpha$. Therefore $\theta_0=\alpha$, $\theta_1=\alpha^{14}$ is a solution of $\frac{\alpha}{\theta_0}+\frac{\alpha^6}{\theta_1}=\alpha^{12}$

References

(1) MacWilliams, F.J. and Sloane,N.J.A.(1977), The Theory of Error-Correcting Codes. North-Holland.

(2) Gorenstein,D.C., Peterson,W.W., and Ziorler, N.(1960), Two-Error-Correcting Bose-Chaudhuri Codes are quasi-perfect. Information and Control 3, pp291-294.

(3) Hartmann, C.R.P., A note on the decoding of double-error-correcting binary BCH codes of primitive length. IEEE Trans. Infor. Theory. 17(1971) pp 765-766.

(4) Moreno, O., Goppa codes related quasi-perfect double-error-correcting codes. IEEE International Symposium on Information Theory, Santa Monica, USA. 1981.

(5) Feng, G.L. and Tzeng, K.K., on quasi-perfect property of double-error-correcting Goppa codes and their complete decoding. Information and Control vol. 61 No. 2, May 1984 pp 132-146.

(6) Chen, C.L. Formulas for the solutions of quadratic equations over GF(2^m). IEEE Trans. Infor. Theory. 28(1982) pp 792-794.

ON THE NUMBER OF DIVISORS OF A POLYNOMIAL OVER GF(2)

Ph. Piret

Philips Research Laboratory
2, av. Van Becelaere B-1170 Brussels, Belgium

ABSTRACT

An upper bound is obtained on the number of polynomials over GF(2) that divide a polynomial of degree n over GF(2). This bound is the solution of a maximisation problem under constraints. It is used to show that most binary shortened cyclic codes (irreducible or not) satisfy the Gilbert bound.

1. INTRODUCTION

Let F be a field and let F[X] be the set of polynomials over F. An element a(X) of F[X] having degree n will be called irreducible if it cannot be written as

$$a(X) = b(X)c(X)$$

where b(X) and c(X) are elements in F[X] of degree at least equal to 1. If F is the real field R, the degree of any irreducible polynomial is at most equal to 2 and there are infinitely many irreducible polynomials of degree 1 and 2. As a consequence, a polynomial a(X) of degree n over R may have as many as 2^n different divisors that are themselves polynomials over R. The situation is different for polynomials over finite fields. In this case the number of irreducible polynomials of any degree is finite but there are irreducible polynomials of any degree. Defining $\psi(i)$ to be the number of irreducible polynomials of degree i over GF(2), we obtain from [1, ch. III] :

$$(2^i-2.2^{i/2})/i \leqslant \psi(i) \leqslant 2^i/i \ .$$

In particular $\psi(1)=2$ and $\psi(2)=1$ hold. As a consequence, a polynomial a(X) of degree n over GF(2) will have (for $n \geqslant 3$) strictly less than 2^n polynomial divisors over GF(2). In this paper we obtain a realistic upper bound on the number of divisors of a polynomial of degree n over GF(2). This bound is obtained in section 2 as the solution to a maximisation problem under constraints. In section 3, we use it to obtain results on the minimum distance of general shortened binary cyclic codes. Kasami [2] has shown that shortened cyclic codes generated by *irreducible* polynomials satisfy the Gilbert bound. His proof implies that *most* irreducible shortened cyclic codes satisfy the Gilbert bound. We generalize this result by showing that *most* shortened cyclic codes (irreducible or not) also satisfy the Gilbert bound. Moreover we show that the mean weight distribution of these codes is "close" to the binomial distribution.

In the sequel we denote by $\lfloor x \rfloor$ and $\lceil x \rceil$ respectively the largest integer $\leqslant x$ and the smallest integer $\geqslant x$.

2. AN UPPER BOUND ON THE NUMBER OF DIVISORS OF A POLYNOMIAL OVER GF(2).

Let a(X) be a polynomial of degree n over GF(2) and express it as the product

$$a(X) = \prod_i [p_j(X)]^{e(j)} \tag{1}$$

where the $p_j(X)$ are the irreducible polynomials over GF(2) and the e(j) are their exponents in a(X). Of course only a finite number of e(j) are nonzero. Defining P(i) to be the set of integers j s.t. the degree of $p_j(X)$ is i, we write

$$a_i(X) = \prod_{j\in P(i)} [p_j(X)]^{e(j)} \tag{2}$$

$$a(X) = \prod_{i=1}^{s} a_i(X) \tag{3}$$

where s is any upper bound on the maximum degree of the irreducible factors of a(X). The number of distinct divisors of a(X) will be denoted by M[a(X)] and the decompositions (1) and (2) lead to

$$M[a(X)] = \prod_j [1+e(j)] \tag{4}$$

$$M[a_i(X)] = \prod_{j\in P(i)} [1+e(j)] \tag{5}$$

We denote by M_n the maximum value of M[a(X)] for all a(X) of degree n and we call maximal any a(X) of degree n satisfying $M[a(X)]= M_n$. To obtain an upper bound on M_n we first prove two properties of maximal polynomials.

<u>Theorem 2.1</u>. Write a maximal polynomial a(X) of degree n as in (3) and suppose that for some i, at least one irreducible polynomial p(X) of degree i appears in (2) with exponent zero. In this case $a_\ell(X)=1$ holds for all $\ell \geqslant i+1$.

<u>Proof</u>. If for some $\ell \geqslant i+1$, $a_\ell(X)$ is not equal to 1, it is divisible by at least one irreducible polynomial q(X) of degree ℓ. In this case $a^*(X)=a(X)p(X)/q(X)$ is a polynomial of degree $n-\ell+i$ ($<n$) satisfying

$$M[a^*(X)] \geqslant M[a(X)]$$

as it follows from (4). Multiplying a*(X) by an arbitrary monic polynomial b(X) of degree $(\ell-i)$ we obtain a polynomial b(X)a*(X) of degree n that satisfies

$$M[b(X)\ a^*(X)] > M[a(X)]$$

so that the polynomial a(X) would not be maximal. Hence the contradiction that proves the theorem. □

Let us denote by n(i) the degree of $a_i(X)$ and by $\psi(i)$ the number of irreducible polynomials of degree i over GF(2). We have:

$$i \sum_{j \in P(i)} e(j) = n(i) . \tag{6}$$

If a(X) is a maximal polynomial it follows from Theorem 2.1 that for some integer s, we have

$$n(i) \geq i\psi(i) , \quad 1 \leq i \leq s-1 \tag{7a}$$

$$n(i) = 0 , \quad i \geq s+1 . \tag{7b}$$

We give now an upper bound on this integer s. This bound will be used in the expression (3) of a maximal polynomial a(X).

<u>Theorem 2.2.</u> Let s be the largest degree of the irreducible factors of a maximal polynomial a(X) of degree n. Then, $\log_2 n \geq s-1$ holds.

<u>Proof</u>: From (7) we obtain

$$n \geq \sum_{i=1}^{s-1} i\psi(i) . \tag{8}$$

Using $\sum_{j|m} j\psi(j) = 2^m$, we obtain 2^{s-1} as a lower bound to the right member of (8). □

We can thus write a(X) as in (3) with $s=1+\lfloor \log_2 n \rfloor$. Let us now consider the expression (5) of $M[a_i(X)]$. We remark that for fixed i, the integers e(j) satisfy :

$$i \sum_{j \in P(i)} [1+e(j)] \leq i\psi(i)+n(i) . \tag{9}$$

$M[a_i(X)]$ is thus upper bounded by the maximum value of the product appearing in (5) under the constraint that the sum of its factors satisfies (9). This maximum value is reached when all factors are equal, i.e. when

$$1+e(j) = [i\psi(i) + n(i)]/i\psi(i) , \quad \text{all } j \in P(i) \tag{10}$$

holds. Since P(i) contains $\psi(i)$ polynomials, this implies:

$$M[a_i(X)] \leq [1+n(i)/i\psi(i)]^{\psi(i)} \tag{11}$$

$$M[a(X)] \leq \prod_{i=1}^{s} [1+n(i)/i\psi(i)]^{\psi(i)} \tag{12}$$

where the bounds can be reached only if all their factors are integers. The inequality (12) gives an upper bound on $M[a(X)]$ when the degrees $n(i)$ of the factors $a_i(X)$ of $a(X)$ are fixed. We modify it to get a general upper bound on M_n that is independent of these $n(i)$. The problem is to maximise the right member of (12) with respect to the indeterminates $n(i)$ under the constraints

$$\sum_{i=1}^{s} n(i) = n . \tag{13}$$

We remark that since ln x the natural logarithm of x in an increasing function of x for $x > 0$, the maximisation of $M[a(X)]$ is equivalent to the maximisation of $\ln M[a(X)]$. To find this maximum we introduce a Lagrange multiplier λ and we define:

$$f[n(1),\dots,n(s),\lambda]= \sum_{i=1}^{s} \psi(i)\ln[1+n(i)/i\psi(i)] -\lambda[\sum_{i=1}^{s} n(i)-n]. \tag{14}$$

We have to solve the system

$$\delta f/\delta n(i)=0 , \qquad 1 \leqslant i \leqslant s \tag{15a}$$

$$\delta f/\delta\lambda = 0 , \tag{15b}$$

that is equivalent to

$$\psi(i)/[i\psi(i)]+ n(i) = \lambda, \qquad 1 \leqslant i \leqslant s \tag{16a}$$

$$\sum_{i=1}^{s} n(i) = n . \tag{16b}$$

The solution of (16) is given by:

$$\psi(i)/[i\psi(i)+n(i)] = \sum_{j=1}^{s} \psi(j)/[n+ \sum_{j=1}^{s} j\psi(j)] = \lambda, \qquad 1\leqslant i\leqslant s \tag{17}$$

Introducing these values in (12) we obtain the bound:

$$\ln M_n \leqslant \sum_{i=1}^{s} \psi(i)\ln\{[n+ \sum_{j=1}^{s} j\psi(j)]/i \sum_{j=1}^{s} \psi(j)\} \tag{18}$$

In this maximisation we could add the constraints

$$n(i) \geqslant 0 \qquad 1 \leqslant i \leqslant s \tag{19}$$

but they are always satisfied by the $n(i)$ given by (17). (We omit the proof).

We give an example to illustrate how good is the bound (18). Choosing n=201 so that s is equal to 8, and using in (18) the values of $\psi(j)$ given in [1], we obtain

$$\ln M_{201} \leqslant 29.44 \tag{20}$$

In other words a polynomial of degree 201 over GF(2) has certainly less than 6.2 10^{12} different polynomial divisors over GF(2). To have an idea of the tightness of this bound we construct a polynomial a(X) as follows. Let $p_r^{(i)}(X)$, $1 \leqslant r \leqslant \psi(i)$, be the rth irreducible polynomial of degree i, define

$$a_i(X) = \prod_{r=1}^{\psi(i)} [p_r^{(i)}(X)]^{e(i)} , \quad 1 \leqslant i \leqslant 6 \tag{21a}$$

$$a_7(x) = \prod_{r=1}^{7} [p_r^{(7)}(X)] \tag{21b}$$

with e(1)=9, e(2)=4, e(3)=3, e(4)=2, e(5)=e(6)=1 and let a(X) be $\Pi_{i=1}^{7} a_i(X)$. M[a(X)] is computed by use of (4):

$$M[a(X)] = 10^2.5.4^2.3^3.2^6.2^9.2^7 .$$

This quantity is approximately equal to 0.906 10^{12} and is to be compared with the bound 6.2 10^{12} obtained above.

For asymptotic evaluation we modify (17) as follows. First we need the result

$$\sum_{i=1}^{s} \psi(i) \leqslant [2^{s+1}/(s+1)](1+o(1)) \tag{22}$$

obtained in the Appendix. Second we upper bound $\sum_{j=1}^{s} j\psi(j)$ by 2^{s+1} and we note that with $s=1+\lfloor \log_2 n \rfloor$, we have $2^{s-1} \leqslant n \leqslant 2^s$. As a consequence

$$[n+ \sum_{j=1}^{s} j\psi(j)]/ \sum_{j=1}^{s} \psi(j) \leqslant (1+n/2^{s+1})(s+1)(1+o(1))=\alpha(s+1)(1+o(1))$$

holds, where $\alpha=(1+n/2^{s+1})$ is in the interval [5/4, 3/2]. From (17) we obtain:

$$\ln M_n \leqslant \sum_{i=1}^{s} \psi(i) \ln[\alpha(s+1)(1+o(1))/i] . \tag{23}$$

We split this sum into $\sum_{i=1}^{s-r-1} + \sum_{i=s-r}^{s}$ with $r=\lceil s^{1/2} \rceil$. From (22) these two sums satisfy:

$$\sum_{i=1}^{s-r-1} \psi(i)\ln[\alpha(s+1)/i] \leqslant 2^{s-r}(1+o(1))\ln[\alpha(s+1)]/(s-r) ,$$

$$\sum_{i=s-r}^{s} \psi(i)\ln[\alpha(s+1)/i] \leqslant 2^{s+1}(1+o(1))\ln[\alpha(s+1)/(s-r)]/(s+1) ,$$

so that for large s, (23) becomes:

$$\ln M_n \leqslant [2^{s+1}/(s+1)](1+o(1))\ln \alpha$$

or equivalently

$$M_n \leqslant \alpha^{[2^{s+1}/(s+1)](1+o(1))}$$

with $\alpha \in [5/4, 3/2]$.

Using $s \leqslant 1+\log_2 n$ and $\beta = \alpha^4 \leqslant 5.0625$, we write this inequality as

$$M_n \leqslant \beta^{\frac{n}{2+\log_2 n}[1+o(1)]} \quad . \tag{24}$$

In the next section we shall need the property that $n^{-1}\ln M_n$ goes to zero for large n. This is made evident by (24).

3. MINIMUM DISTANCE AND WEIGHT DISTRIBUTION OF SHORTENED CYCLIC CODES.

We consider a family $\mathcal{C}(n,R)$ of N linear codes of length n and dimension $k=\lceil Rn \rceil$ over GF(2), having the property that any nonzero word appears in at most M of these codes. By $A_C(w)$, we denote the number of codewords of weight w in the code C and by $\bar{A}(w)$ we denote the mean value of $A_C(w)$ for $C \in \mathcal{C}(n,R)$. The total number $N\,\bar{A}(w)$ of codewords of weight w in the N codes of $\mathcal{C}$ is at most $\binom{n}{w}M$ which implies

$$\bar{A}(w) \leqslant \binom{n}{w}\ M/N \tag{25}$$

Let us define:

$$B_C(w) = \sum_{i=1}^{w-1} A_C(i)$$

$$\bar{B}(w) = \sum_{i=1}^{w-1} \bar{A}(i)$$

$$S(w) = \sum_{i=1}^{w-1} \binom{n}{i}$$

The total number $N\bar{B}(w)$ of nonzero codewords of weight $\leqslant w-1$ in the N codes of $\mathcal{C}$ is at most equal to $S(w)M$ which implies

$$\bar{B}(w) \leqslant S(w)\ M/N\ . \tag{26}$$

Let us particularise $\mathcal{C}$ to the family of $(n,k=\lceil Rn \rceil)$ shortened cyclic codes. To any polynomial $g(X)=1+\sum_{i=1}^{n-k-1} g_i X^i + X^{n-k}$ over GF(2), we associate the binary code

$C(g)$ of length n and dimension k:

$$C(g)=\{a(X)g(X) \ : \ \deg a(X) \leqslant k-1\}$$

where deg 0, the degree of the zero polynomial is assumed to be zero. The set $C(n,R)$ of these shortened cyclic codes $C(g)$ thus contains $N=2^{n-k-1}$ different codes. Some years ago, Kasami [2] has considered the class of irreducible shortened cyclic code, i.e. the class of codes $C(g)$ for which the polynomial $g(X)$ is irreducible, and he has proved that this subclass $C^*(n,R)$ of $C(n,R)$ satisfies the Gilbert bound [3]. So the *existence* of shortened cyclic codes near the Gilbert bound is already known and moreover it is not difficult to prove that most codes in $C^*(n,R)$ are close to the Gilbert bound. However the number of codes in $C(n,R)$ is approximately n-k times as large as the number of codes in $C^*(n,R)$, [1] . To prove that *most* codes in $C(n,R)$ are close to the Gilbert bound, we need a specific argument given in the sequel. As the reader may check, this will make the class $C(n,R)$ suitable as a class of inner codes for the Justesen construction [4,5] since it is no more necessary to check the polynomials $g(X)$ for irreducibility. The argument will be to evaluate for fixed n and for $k=\lceil Rn\rceil$ an upper bound on the parameter $\bar{B}(w)$ of $C(n,R)$. With $N=2^{n-k-1}$ we obtain from (24),(26) and Stirling approximation that for fixed $\omega \in]0,1/2[$

$$n^{-1} \log_2 \bar{B}(\lceil \omega n\rceil) \leqslant H(\omega)-(1-R)+o(1) \tag{27}$$

holds with $H(x)= -x \log_2 x -(1-x) \log_2(1-x)$. In the same way we obtain from (24) and (25)

$$n^{-1} \log_2 \bar{A}(\lceil \omega n\rceil) \leqslant H(\omega)-(1-R)+o(1) \tag{28}$$

for all $\omega \in]0,1[$. By use of Chebyshev's inequality [6], it follows from (27) that the fraction $\psi(\delta)$ of codes in $C(n,R)$ having a minimum distance strictly smaller than $d=\delta n$, satisfies:

$$\psi(\delta) \leqslant 2^{n[H(\delta)-(1-R)+o(1)]}$$

For fixed R and large n, it becomes vanishingly small if $H(\delta) < (1-R)$ holds, i.e. if δ is smaller than the Gilbert bound. In the same way, it follows from (28) that the fraction $\phi(\varepsilon,\omega)$ of codes in $C(n,R)$ having $A(\lceil \omega n\rceil) \geqslant 2^{n[H(\omega)-(1-R)+\varepsilon]}$ satisfies:

$$\phi(\varepsilon,\omega) \leqslant 2^{-n[\varepsilon-o(1)]} , \qquad \text{all } \omega \in]0,1[\ .$$

APPENDIX

To prove (22) we use the bound

$$\psi(i) \leqslant 2^i/i ,$$

and we split $\sum_{i=1}^{s}$ into $\sum_{i=1}^{r} + \sum_{i=r+1}^{s}$. The first of these two sums is bounded by

$$\sum_{i=1}^{r} \psi(i) \leqslant \sum_{i=1}^{r} 2^i < 2^{r+1}$$

and the second one is bounded by

$$\sum_{i=r+1}^{s} \psi(i) \leqslant \Big(\sum_{i=r+1}^{s} 2^i\Big)/(r+1) < 2^{s+1}/(r+1)$$

These bounds lead to

$$\sum_{i=1}^{s} \psi(i) \leqslant [2^{s+1}/(s+1)][(s+1)/(r+1)+(s+1)\, 2^{r-s}] . \tag{A1}$$

Choosing $r= \lceil s-s^{1/2} \rceil$ we obtain from (A1) :

$$\sum_{i=1}^{s} \psi(i) \leqslant [2^{s+1}/(s+1)][1+o(1)]. \tag{A2}$$

REFERENCES

[1] S.W. GOLOMB, *Shift Register Sequences*, Holden-Day, San Francisco 1967.

[2] T. KASAMI, "An upper bound on k/n for affine-invariant codes with fixed d/n", *IEEE Trans. Inform. Theory*, vol. IT-15, pp. 174-176, January 1969.

[3] F.J. MACWILLIAMS and N.J.A. SLOANE, *The Theory of Error Correcting Codes*, North-Holland, 1977.

[4] J. JUSTESEN, "A class of constructive asymptotically good algebraic codes", *IEEE Trans. Inform. Theory*, vol. IT-18, pp. 652-656, September 1972.

[5] E.J. WELDON, "Justesen's construction-The low-rate case", *IEEE Trans. Inform. Theory*, vol. IT-10, pp. 711-713, September 1973.

[6] R.G. GALLAGER, *Information Theory and Reliable Communication*, Wiley, New York, 1968.

Multivariate Polynomial Factoring and Detection of true factors

Denis LUGIEZ
LIFIA, Grenoble BP 68
38402 Saint Martin d'Hères, France

1 Introduction

Nowaday, the problem of factoring polynomials is known to be a polynomial-time problem, due to the work of Lenstra [LE 82], but his algorithms are not efficient enough to supersede the classical ones. A lot of work is done to improve them, but from a pratical point of view, it is still worthy to improve the classical algorithm, and this work belongs to this kind of improvements. The polynomials to factor are dense polynomials, the case of sparse polynomial has been treated by Zippel and some special cases are examined in Davenport [DA 84]. In univariate polynomial factorization, the early detection of the true factors usually spares a lot of computation time. Here a refinement of the lifting process presented in [LU 84], based on Partial Fraction Decomposition as suggested by Viry in [VI 80], is given and discussed. This lifting allows a test for the early detection of the true factors as in the univariate case. Moreover, the reduction of the number of lifting steps reduces the worst-case complexity of the reconstruction step and reduces the time required by the factorization process. The method to compute the homogeneous parts have been refined, and a new one based on the Taylor's expansion is presented. Some measurements have been done to compare this lifting with the usual one implemented in the Computer Algebra Sytem SAC 2/ ALDES and have shown that it is faster, although the test for the early true factor detection is not yet implemented.

2 Presentation of the algorithm

2.1 Notations.

The polynomials to factor are multivariate squarefree primitive polynomials over the integers. This means that the GCD of the coefficients equals 1, and that there is no multiple factor. The polynomials are dense polynomials, but in general the factors may not be dense polynomials. $P(X, X_1, ..., X_m)$ is a m+1 variable primitive squarefree polynomial to factor over the integers.

M is a bound over the maximum of the absolute value of the integer coefficient of any factor of the polynomial. In SAC2, the Gelfond's bound is used.

$a_1, ..., a_m$ are integers which are substituted in the variables X_i in the first step of the factorization.

I_i is the ideal of $Z[X][X_1, ..., X_m]$ generated by the homogeneous polynomials of degree i in the new variables $Y_i = X_i - a_i$.

p, the global degree of the polynomial in $X_1, ..., X_m$ is the maximum of the global degree of each monomial. The global degree of a monomial is the sum of the partial degree in $X_1, ..., X_m$. The global degree of the leading coefficient of P in the $X_1, ..., X_m$ is denoted by p'. These degree remain unchanged after the change of variable $Y_i = X_i - a_i$.

The degree of$P$ in Y_i which is also the degree in X_i is n_i. The degree in the main variable X is n.

Each polynomial of global degree p is equal to the sum of its homogeneous components of degree i in the variables $Y_1, ..., Y_m$, from i=0 to p. These components are called homogeneous parts in the following and denoted P_i.

example : If P(X,X_1,X_2) $= XX_1X_2 + X_1X_2^2 + X^3X_1$ and $a_1 = 1, a_2 = -1$

then p = 3
$P_0 = X^3 - X + 1,$
$P_1 = X^3(X_1 - 1) + X(X_2 - X_1 + 2) + X_1 - 2X_2 - 3,$
$P_2 = X(X_1X_2 - X_2 + X_1 - 1) - 2X_1X_2 - 2X_1 + X_2^2 + 4X_2 + 3,$
$P_3 = X_2^2X_1 - X_2^2 + 2X_1X_2 - 2X_2 + X_1 - 1$

2.2 Scheme of the factorization process

The proposed method has the same steps than the classical one :

1. Choose suitable integers a_i and factor $P(X, a_1, ..., a_m)$ over Z[X].
2. Compute a suitable bound M over the coefficients of the factors, and lift the factorisation over the integers which gives a factorization mod(I_1, M) to a factorization mod$(I_{(p+p')/2+1}, M)$
3. Reconstruct the true factors over the integers from the factors mod $(I_{(p+p')/2+1}, M)$

Three main problems appear in this factorization process :

1. The extraneous factors problem : when a irreducible factor of P splits into more than one factor after the evaluation in the $a_i's$
2. The leading coefficient problem : when the polynomial P is not monic, i.e. its leading coefficient is not 1.
3. The bad zero problem : when the evaluation points a_i are not zero.

The two first problems are difficult to solve in theory, although a theorem of Hilbert can be used for the solution of the first one, but in practice heuristics solutions exist and usually overcome the difficulty. The extraneous factor problem is solved by testing several factorizations $P(X, a_1, ..., a_m)$ for several choices of the a_i's. The second one is solved by trying to find the leading coefficient of a factor from its integer image after the evaluation in the a_i. If this heuristic succeeds, the the knowledge of the factors mod $I_{p/2}$ is sufficient. See Wang [WA 78] for a complete presentation of these solutions. Another method, implicitely adopted above, is to force the leading coefficient of the factor to be the leading coefficient of the polynomial and to consider the factor as a factor of the product of the polynomial by its leading coefficient. The last problem which appears when the a_i's are not equal to zero requires more attention and is discussed in section 3.

From now, we suppose that no extraneous factor appears and that the leading coefficient of each factor is known unless the contrary is explicitely assumed. This allows to replace $(p + p')/2$ *by* $p/2$ *in the preceeding scheme.*

Here p/2 means the floor of p/2. One may already remark that the usual liftings compute a factorization mod $(Y_1^{n_1+1}, ..., Y_m^{n_m+1})$ with n_i the degree of P in Y_i and $p \leq \sum_{i=1}^{n} n_i$.

2.3 The Lifting Process

The lifting process is briefly described, a complete description is found in [LU 84]. For the sake of simplicity, a serial one is given but a parallel one exists.

Partial Fraction decomposition Lifting

P is the polynomial to factor, P_0 is its univariate image after the evaluation in the $a_i^{\prime}s$, F_i^0 for i=1, 2 are the factors of the univariate image.

1. Compute a bound M of the coefficient of the factors such that M=q^k with q a prime number which does not divide the leading coefficient of P_0.
2. Compute the partial fraction decomposition of $X^i/F_1^0F_2^0$ in $Z_M[X]$ for i=0,...,n the degree of P in X.
3. $F_1 = F_1 + f_1^i; F_2 = F_2 + f_2^i$ where f_j^i is the homogeneous part of degree i of the leading coefficient of the factor F_i.
4. For $i = 1, ..., p/2$ do
   begin
   - Compute P_i the homogeneous part of degree i of P.
   - $$S_i = P_i - f_1^iF_2^0 - f_2^iF_1^0 - \sum_{l+m=i,l\neq 0,m\neq 0} F_1^lF_2^m$$
   - Compute the partial fraction decomposition of $S_i/F_1^0F_2^0 = F_1^i/F_1^0 + F_2^i/F_2^0$
   - $F_1 = F_1 + F_1^i; F_2 = F_2 + F_2^i$

   end

end

When the leading coefficients of the factors are not known, the first one f_1 is set to 1, and the second one is set to A_n the leading coefficient of $P(X, X_1, ..., X_m)$. Moreover the factors are computed mod $I_{p+p'/2}$. Then, one gets the monic factors of $P(X, X_1, ..., X_m)$ mod$(I_{(p+p')/2+1}, M)$, except the last one which is multiplied by A_n. When more than two factors are to be lifted, the algorithm is performed recursively.

3 Computation of the homogeneous parts

3.1 The three methods

To perform the lifting, one must be able to compute the homogeneous part of the polynomial $P(X, X_1, ..., X_m)$ in the variables Y_i. Three methods are described here. The first one is to realize the change of variable Y_i=$X_i - a_i$ and to perform the entire lifting with these new variables and then to compute the homogenous parts by picking up the monomials of degree i in the polynomial in Y_i. Finally the inverse change of variables is done when the factors are found. This method is exponential.

The second one is to use an improvement of the method of [LU 84] which uses Euler's identity. The homogeneous parts are computed in the old variables X_i and no change of variables is needed. It is briefly described here :

Compute the sequence G_i for i=0,...,p such that :
$G_0 = P$,
$G_{i+1} = (p-i)G_i - \sum_{l=1}^{m} DG_i/DX_l$

Then these G_i satisfy the triangular linear system :

$G_1 = a_{1,1}P_1 + ... + a_{1,j}P_j + ... + a_{1,p}P_p$

.

.

$G_{p-j} = a_{j,1}P_1 + ... + a_{j,j}P_j$

.

$G_p = a_{p,1}P_1$

where the $a'_{i,j}s$ are known integers.

This system is easily solved to get the $P_i's$, and it is easy to compute them only for i=1,...,p/2. The proof of the method is similar to the proof given in [LU 84] and relies upon Euler's identity.

The third one is to use Taylor's expansion. The Taylor's expansion has been used by Wang [WA 78] in the classical case, and we will show that it is possible to use it in a similar way in the proposed lifting.

Taylor's formula : $P(X) = P(a) + (X-a)D^1(a) + ... + 1/n!(X-a)^n D^n(a) \bmod (X-a)^{n+1}$ where P is a polynomial of $Z[X_1, ..., X_m][X]$ and the D^i denote the ith partial derivative of P with respect to X.

Since the homogeneous parts P_i are the sum of the products of degree i in the $Y_i = X_i - a_i$, the Taylor's expansion up to degree $p/2$ can be used recursively to compute these $P_i's$. A recursive computation of the $1/i!D^i$ avoids the costly computation of i! and allows a computation of any P_i.

The algorithm for computing the homogeneous parts of a polynomial is briefly described.

P is a polynomial in m+1 variables $X, X_1, ..., X_m$ but it is taken in a polynomial in the variables $X_1, ..., X_m$. The algorithm computes the homogeneous parts of P in $Y_1, ..., Y_m$ for i=0,...,m.

$$S_p = FUNCTION(m, p, P)$$

S_p is the homogeneous part of degree p of the Polynomial P, m is the number of variables, p is the degree of the homogeneous part to be computed.

1. $A_0 = P$; $B_0 = P(X, a_1, ..., a_m)$;
2. For i=0,...,p do ($A_{i+1} = 1/(i+1)DA_i/DX_m$; $B_{i+1} = A_{i+1}(a_m)$);
3. For i=0,...,p do $S_p = S_p + Y_m^i.FUNCTION(m-1, p-i, B_i)$

3.2 Comparison

The first method has the disadvantage to change sparse polynomials in dense polynomials and is exponential. The second one avoids the change of variables but deals with homogeneous polynomials in Y_i expressed in the X_i's . These polynomials have few terms when expressed in the Y_i but are almost always dense when expressed in the X_i's. The third method avoid the expansion of terms and deals with homogeneous polynomials expressed in the Y_i's. Since the timing measurements for the second method revealed that it was too bad, the comparisons were restricted to the first method, the change of variables, and the third one, the Taylor's expansion. The comparisons were achieved with the ALDES/SAC2 system running on a Vax 750 under UNIX 4.2 BSD.

The m-uple T is the set of the a_i's. T_t is the time required for the computation of the homogeneous parts of the polynomial from 0 to $p/2$ using the Taylor's expansion.
T_h is the time required by the computations of the homogeneous parts of the polynomial from 0 to $p/2$ with the change of variables performed by Horner's rule.
All times are given in milliseconds.

$$P(X,X_1,X_2) = (X + X_1 + X_2^4)(X^2 + XX_1^2 + X_2^2 + 1)$$

T=(1,1)
T_h = 1310 MS
T_t = 630 MS

$$P(X,X_1,X_2) = (X^2 - XX_1^2 + X_2^4)(X^2 + 2XX_2 + X_1X_2^2)$$

T=(2,1)
T_h= 3880 MS
T_t= 650 MS

$$P(X,X_1,X_2,X_3) = (45X^3 - X_1^2 - 9X_1^3 + 3X_2^3 + 2X_2X_3)(47XX_1 + X_2^3X_3^2 - X_3^2)$$

T=(1,2,3)
T_h =9850 MS
T_t= 1480 MS

$$P(X,X_1,X_2) = (X^2 - X_1^{10})(X^2 - X_2^{10})$$

T = (2,1)
T_h = 23530 MS
T_t = 1400 MS

$$P(X,X_1X_2) = (X^2 - X_1^{20})(X^2 - X_2^{20})$$

T=(2,1)
T_h = *
T_T = 9130 MS

* means that the program ran out of memory.

These examples show that the Taylor's expansion method is better than the method based on the expansion of the polynomial. This is due to the fact that the expansion of terms is ocstly and that many useless computations are performed, since the polynomial is completely expanded, which means that the terms up to degree p are computed. On the contrary, the Taylor's expansion based method computes only what is really needed, i.e the terms up to degree p/2.

4 Reduction of the complexity and Detection of the factors

4.1 Reduction of the complexity

The complexity of the proposed lifting is reduced by the fact that the lifting is stopped after p/2 iterations and not after p iterations as in the lifting described in [LU 84]. One have to remark that the classical method stops after $\sum_{i=1}^{m} n_i$ iterations, which is greater or equal p. It is possible to stop the lifting when the factors are computed mod $I_{p/2}$ because of the following proposition.

Prop : *The product of a polynomial of $Z[X][X_1, ..., X_m]$ of degree p by a polynomial of $Z[X][X_1, ..., X_m]$ of global degree q has a global degree equal to p+q.*

The proof is trivial, it may help to consider an ordering on the monomials of the same degree. Then, either a factor of P has a degree less than $p/2$ or its cofactor has a degree less than $p/2$; and the knowledge of the polynomials mod$I_{p/2+1}$ is enough provided both the factor and its cofactor are tested for trial division in the step of the reconstruction of the factors.

An example shows the reduction of the number of step when compared with the classical method. Anyway, the new one has less iterations to perform than the classical one since $p/2 < p \leq \sum_{i=1}^{m} n_i$

Example : $P(X, X_1, X_2) = (X + X_1^4 + 1)(X + X_1^2 + X_2^3)$
then p=7, n_1=6, n_2=3, hence the classical lifting performs $n_1 + n_2 = 9$ iterations and the proposed one performs 3 iterations.

It has been shown for the univariate lifting that the cost of one step exceeds the cost of all the previous ones and this remains true for the multivariate lifting when the polynomials are dense. Therefore, it is of a very importance for the performance of the algorithm to avoid the last steps.

If there are some extraneous factors or if the leading coefficients of the factors can not be found, one must lift the factors up to $I_{(p+p')/2}$ and perform an exponential search to find the true factors. The fact that the factors are known mod$I_{p/2}$ rather than modI_p spares a lot of computational time, when these modular factors are dense. The experience has shown that it is the case specially with the bad zero problem or the leading coefficient problem, altough one could build exotic counterexamples.

When the modular factors are computed mod$I_{p/2}$, it is necessary to test for trial division both a factor and its cofactor. If they were known modI_p, a single trial division is enough. In the worst case, i.e. a irreducible polynomial P, one tests all the 2^r possible modular factors, in place of 2^{r-1} . But the computations are done in the space of the homogeneous polynomials of degree less or equal p+p'/2 which has the dimension $C^m_{m+(p+p')/2}$ and not in the space of the homogeneous polynomials of degree less or equal p+p'/2 which has dimension $C^m_{m+p+p'}$ where C^p_n denotes the binomial coefficient. Since the polynomials are dense and the second dimension is larger than the first one, the complexity is decreased.

4.2 Early detection of true factors

The early detection of the factors means that a true factor of P is found during the lifting before p/2 iterations have been performed. It relies on the fact that if P has r factors then one of the factors has a degree $\leq p/r$. Then if no extraneous factor exist, a heuristic test performed after p/r iteration will find it. This test can not be done in the classical lifting, because the factors are lifted variable after variables. However, Wang in [WA 84], suggests a partial test for the partial factors. This test may be also implemented in our scheme.

The test for finding a true factor : When p/r iterations have been performed, the currently lifted factor is tested to see if it is equal to a true factor of the polynomial. If the test is succesfull, the polynomial being factored is replaced by its quotient by this factor. Therefore the degree p and the bound M are decreased, yielding a sparing of computations for the remaining factors.

The partial test for true factor : One chooses the lowest degree variable e.g X_{i_0} and test after n_{i_0} iteration if one can get a factor of $P(X, a_1, .., a_{i_0}, X_{i_0}, a_{i_0+1}, ..., a_m)$ from the factor mod $I_{n_{i_0}}$. This test seems to be implemented by Wang, in his factorization module. But this test helps more for detecting a extraneous factor than to compute the factor. Instead of giving up the current factorization as Wang does when this test reveals that an extraneous factor is in the factorization, it could be better to make products of these factors in order to get a modular factor coming from a true factor. Some refinements may be used to deal with this case, specially when there are few factors.

With respect to the detection of the factors, a parallel algorithm is better than a serial one since all the factors are lifted together, and in the serial lifting the factor currently lifted may have a degree greater than $p/2$ or even $p/2$ which forbids the early detection of this factor.

5 Comparison

5.1 Representation of Polynomials

The most widely used polynomial representation is a recursive one, i.e each polynomial is a polynomial in a main variable with coefficients in a ring, these coefficients may be also polynomials. This representation is ill-suited when there the variables play the same role, and when one wants to exchange the main variable with an other one. This representation does not fit well our algorithms which consider homogeneous polynomials. This problem also appears in Grobner Basis computations. Nowaday, if one would use a different representation for polynomials, he has to implement it and all the related basic operations. Some packages for Grobner basis computations in SAC2 and REDUCE, provide a distributive representation of polynomials, but they are not currently available. The problem of the choice of representation of the objects is a fundamental one and has to be investigated at the conception stage of a Computer Algebra System.

5.2 Measurements

The measurements were performed with ALDES/SAC2 of COLLINS and LOOS, running on a Vax 750 under UNIX. The memory cell allocation was set to 200000 using the Pragma TNU assignement.

The new algorithm has been implemented in a serial way, and the test for an early detection of the true factors is still lacking. Since the classical lifting implemented in SAC2/ ALDES does not support the heuristic which finds the leading coefficient of the factors, the version of our lifting used for the measurements also lacks this feature. But one should remind that this heuristic would save a lot of computations time, due to the stop of the lifting after p/2 iterations in place of p+p'/2 and to the preservation of the sparsity of the polynomials. The tested polynomials were dense, but their factors had about half of their coefficients equal to zero.

T is the time required by the first part of the factorization, i.e to find the a_i and to factor $P(X, a_1, ..., a_m)$.

T_{cl} is the time required by the classical lifting method implemented in SAC2 to compute the factors from the univariate factorization.

T_{new} is the time required by the proposed method to compute the factorization from the univariate factorization.

Monic Polynomials .

rl is the number of variables.

Time	rl = 2	rl =3	rl = 3	rl= 3	rl = 3
T	1859 MS	1604 MS	5394 MS	1440 MS	4017 MS
T_{cl}	1912 MS	4729 MS	31791 MS	1813 MS	43360 MS
T_{new}	583 MS	609 MS	2587 MS	479 MS	5259 MS

Non-Monic Polynomials.

Time	rl=3, p=10	rl=3, p=9	rl =4, p = 14
T	4770 MS	5180 MS	28380 MS
T_{cl}	596790 MS	2807180 MS	*
T_{new}	46830 MS	189270 MS	1598290 MS

* means that the process ran several hours without any result.

5.3 A Factorization Module

We do not intend to describe what a Factorization Module for multivariate polynomials should be, but we point out some features which should be investigated when building such a module.

First of all, the univariate factoring module should be able to handle polynomials with large coefficients since the multivariate factorization yields such polynomials, and it should be fast enough. This may imply the implementation of heuristics and probabilistic algorithms.

A first step in the factorization is to recognize special cases. This is already done in Macsyma for example, but only for few cases, and it could be extended, see [DA 83], and the recognition of special cases in multivariate polynomial factoring is difficult and should be worked out.

Then the kind of the polynomial has to be found, is it sparse or dense? If it is sparse, then the algorithms for sparse polynomials are applied. Since these algorithms are probabilistic, they may fail, in this case, the algorithms for dense polynomials are called.

In case of dense polynomials, the problem should be transformed in order to get the best input. This means for example that the main variable is changed according to considerations about the degrees, the leading coefficients, what algorithm will be used (Lentra's, the classical one, the proposed one,...)

Then the evaluation points are chosen, either the smallest possible ones or at random. The heuristics related to the leading coefficients and the extraneous factors are tried, and they may yield a new choice for the evaluation points.

According to the number of univariate factors, their degree, their sparsity, the lifting algorithm is chosen to be the classical one or the proposed one, serial or parallel, with the early detection of the factors or not. The bound on the coefficient of the factors which is often too large can be changed into a lower one.

Then the lifting is performed, and some factors are found. If each univariate factor raises a multivariate factor or if no heuristic is used, then the factorization is done. Otherwise, the problem has to be solved for the remaining factors either by restarting from scratch or by coming back in the heuristic choices performed.

The overall strategy retained for this module is to use the more heuristics as possible in order to get a good average behaviour, but to use the available algorithms when heuristics fail or do not solve completely the problem.

6 Conclusion

This paper presented a refinement of the lifting process based on partial fraction decomposition and shows how it is possible to reduce the number of steps in this lifting, moreover it advocates for the early detection of the true factors which is intende to save computational time. The measurements have shown that the proposed method, although some heuristics are still lacking, is interesting to use. The issue about parallel or serial lifting is not treated, nor a possible combination of both to the price of a higher memory cost. The representation of polynomials and more generaly, the representation of objects in Computer Algebra Systems is still pending, partly due to the fact that there is not "best" solution. The refinements proposed here may be extended to GCD computations since our method yields a GCD algorithm as the classical lifting has yielded the EEZGCD algorithm. It could be adapted in more general rings, in the same way Lauer [LA 83], has done it for the Hensel Lifting.

References

[DA 83] J.H. DAVENPORT : "Factorization of Sparse Polynomials" EUROCAL 83, London, LNCS 162 p214-223.

[GA 83] J. von zur GATHEN : "Hensel and Newton methods in valuation rings", Technical report, Dpt of Computer Science, university of TORONTO.

[KA 82] E. KALTOFEN : "Factorization of Polynomials", Computer Algebra, Buchberger & al. Editors, Springer Verlag.

[KU TO 77] H.T. TONG , D.M. TONG : " Fast Algorithms for Partial Fraction Decomposition", SIAM J. of Algorithms 6, p582-593, 1977

[LA 83] M. LAUER : "Generalised p-adic construction" , SIAM J. of Computing 12, p395-410, 1983.

[LE 82] A.K. LENSTRA : " Lattices and Factorization of Polynomials" EUROCAM 82, Marseille, LNCS 144 p32-39.

[LU 84] D. LUGIEZ : " A new Lifting Process for Multivariate Polynomial Factorization", EUROSAM 84, Cambridge, LNCS 174 p297-309.

[MO YU 83] J. MOSES, D.Y.YUN : " The EZGCD Algorithm" ACM National Conference 1973, p159-166.

[VI 80] G.VIRY : "Factorisation des Polynomes à plusieurs variables" RAIRO Informatique théorique 14, p209-223, 1980.

[WA 78] P.S. WANG : "An Improved Multivariate Polynomial Factoring Algorithm", Math. of Computation 32, p1215-1231, 1978.

[WA 84] P.S. WANG : "Implementation of a *p-adic* package for polynomial factorization and other related operations", EUROSAM 84, Cambridge, LNCS 174, p86-99

[YU 74] D.Y. YUN : " The Hensel Lemma in Algebraic Manipulation", Ph. D. Thesis, Dpt of Mathematics, MIT 1974.

[ZI 79] R.E. ZIPPEL : "Probabilistic Algorithms for Sparse Polynomials", Ph.D. Thesis, MIT 1979.

DISCRIMINANTS AND THE IRREDUCIBILITY OF A CLASS OF POLYNOMIALS

Oscar Moreno de Ayala
Department of Mathematics
University of Puerto Rico
Río Piedras, Puerto Rico

There has been some interest in finding irreducible polynomial of the type $f(A(x))$ for certain classes of linearized polynomials $A(x)$ (see [1], [2], [3], [6],) over a finite field $GF(P^m)$. The main result of this paper proves the stronger result that there are no further irreducible cases of $f(A(x))$ for an extended class that contains that of linearized polynomials, but for $p = 2$. In order to reach this result, and also of independent interest, the discriminant and the parity of the factors of polynomials $f(A(x))$ is computed.

The discriminant D of a polynomial $A(x) = \prod_{i=1}^{n} (x - \alpha_i)$ is defined $D(A) = \prod\prod_{1 \le i < j \le n} (\alpha_i - \alpha_j)^2$. Then it is not hard to prove (see |4|) that $D(A) = (-1)^{n(n-1)/2} \prod_{i=1}^{n} A'(\alpha_i)$

Let us consider now a polynomial with coefficients in $GF(p^k)$:

$$A(x) = X^{pi} + A_1X^{p(i-1)} + \dots + A_{i-1}X^p + A_iX + A_{i+1}.$$

For $A(x)$ we have the following lemma, where $D(f)$ is computed in $GF(p^k)$, and therefore $p = 0$ (in $GF(p^k)$).

LEMMA 1

For $A(x)$ as defined above we have $D(A) = (-1)^{pi(pi-1)/2}(A_i)^{pi}$.

PROOF: Obvious using the formula for $D(A)$ in terms of A'. An important class of polynomials $A(x)$ contained in the above class for which the lemma applies are the affine polynomials (where every term has degree p^j). From now on we will deal with $p=2$, and we will assume $A(x)$ is such that the degree of every term of degree > 4, is divisible by 8 (This includes the affine polynomials). We will compute the discriminant mod 8 in this case.

LEMMA 2

For $A(x)$ as defined above we have

LEMMA 3

$$D(f(A(X)) = (D(f))^n (D(A))^m$$

PROOF:

$f(A(X)) = \prod_{i=1}^{n} (X-\beta_i)$ and furthermore for every root γ_j of f, $A(\beta_i)=\gamma_j$ for exactly n values of i. Then using again the formula for D in terms of derivatives and denoting $s=(-1)^{nm(nm-1)/2}$

$$D(f(A(X)) = s \prod_{i=1}^{nm} (f(A(X))'(\beta_i) =$$

$$= s \prod_{i=1}^{nm} f'(A(\beta_i)) \prod_{i=1}^{nm} A'(\beta_i)$$

$$= s \prod_{j=1}^{m} (f'(\gamma_j))^m \prod_{i=1}^{nm} (A'(\beta_i) =$$

$$= s' \ (D(f))^n \prod_{i=1}^{nm} A'(\beta_i), \text{ where } s'=s(-1)^{nm(m-1)/2}$$

(notice that $s'(-1)^{nm(n-1)/2} = 1$). But

$\prod_{i=1}^{nm} A'(\beta i) = \prod_{i=1}^{nm} \prod_{A(\beta i)=\gamma_j}^{nm} A'(\beta i)$ and the inner product is iqual to

$(-1)^{n(n-1)/2} D(A(X) + \gamma_j) = (-1)^{n(n-1)/2} D(A)$ and the rest follows easily. Now we will prove

THEOREM 1.

Let n, r_A, r_f be the number of irreducible factors in $GF(2^k)$ of f(A), A, f respectively, and A be as in lemma 3. Then

$$r \equiv nr_f + mr_A + nm(\text{MOD } 2).$$

PROOF:

We will use Berlekamp's generalization for $GF(2^k)$ of the binary version of Stickelberger theorem, which can be found in |4|. This states that for any polynomial $g(X) \varepsilon GF(2^k)[X]$ we have:

$D(g) \equiv B_g^2 + 4B_g'$ (MOD 8) for B_g, $B_g' \ \varepsilon \ GF(2^k)$; and if ℓ is the degree of g, $r_g \equiv \ell$ iff $Tr(\frac{B_g'}{B_g^2}) = 0$ (where $Tr\ (X) = X + X^2 + \ldots + X^{2^{k-1}}$)'

Now for simplicity say $D(f(A)) \equiv B^2 + 4B''$ (MOD 8). Let us remark that

$$(D(g))^i \equiv \begin{cases} B_g^2 & \text{if } i \text{ is even} \\ & (\text{MOD } 8) \\ B_g^{2(i-1)} D(g) & \text{if } i \text{ is odd} \end{cases}$$

$$D(A) \equiv \begin{cases} A_i^n \quad (\text{MOD } 8) & \text{if } n>4 \\ A_2^4 + 4(A_1^2A_2^2 + A_2^4) \quad (\text{MOD } 8) & \text{if } n=4 \end{cases}$$

PROOF:

Now we compute D(A) (MOD 8). We can proceed in a similar manner to Lemma 1 and then it is clear $D(A) \equiv$

$$\equiv \prod_{j=1}^{n} (4A_{i-2}\alpha_j^3 + 2A_{i-1}\alpha_j + A_i) \quad (\text{MOD } 8).$$

Since we are computing MOD 8, it is clear that after cross multiplying the only possible product containing any α_j^3 must have n-1 of the A_i's. In other words $D(A) \equiv$

$$\equiv 4A_{i-2}A_i^{n-1} \sum_{j-1}^{n} \alpha_j^3 + \prod_{j=1}^{n} (2A_{i-1}\alpha_j + A_i) \quad (\text{MOD } 8).$$

In a similar way we find the last product to be:

$$\prod_{j=1}^{n} (2A_{i-1}\alpha_j + A_i) = 4A_{i-1}^2 A_i^{n-2} \sum_{\substack{j,k=1 \\ j \neq k}}^{n} \alpha_j\alpha_k + A_{i-1}A_i^{n-1} \sum_{j=1}^{n} \alpha_j + A_i^n$$

Therefore:

$$D(A) \equiv 4A_{i-2}A_i^{n-1} \sum_{j=1}^{n} \alpha_j^3 + 4A_{j-1}^2 A_i^{n-2} \sum_{\substack{j,k=1 \\ j = k}}^{n} \alpha_j\alpha_\kappa + 2A_{i-1}A_i^{n-1} \sum_{j=1}^{n} \alpha_j + A_i^n$$

Now it is clear that the last two summations are respectively the n-2 and n-1 coefficients of A and they are = 0 if $n \neq 4$. Also using Newton's identities ((9.32 of refence |4|)) we can compute $\sum_{j=1} \alpha_j^3$ in terms of the coefficients of A and find this last summation to be = 0 or $-3A_2$, and the second only if n = 4. Therefore

$$D(A) \equiv \begin{cases} A_i^n \quad (\text{MOD } 8) & \text{if } n > 4 \\ A_2^4 + 4A_1^2A_2^2 + 4A_2^4 \quad (\text{MOD } 8) & \text{if } n = 4 \end{cases}$$

We will deal now with the discriminant of a composition of polynomials f(A(x)) where $A(X) = X^n + A_1X^{n-1} + \ldots + A_{n-1}X + A_n$ and $f(x) = x^m + f_1x^{m-1} + f_{m-1}x + f_m$. The computation of the discriminant will be done (MOD 8).

and also that if $D(f(A)) \equiv B^2 + 4B' \equiv (B^2 + 4B_g')(B_h^2 + 4B_h') \equiv \kappa^2(B_e^2 + 4B_e) \pmod 8$ for $\kappa^2 \in GF(2^k)$ then

$$Tr(\frac{B'}{B^2}) = Tr(\frac{B_g'}{B_g^2}) + Tr(\frac{B_h'}{B_h^2}) = Tr(\frac{B_e'}{B_e^2}).$$

We can say that $r = nm$ iff $Tr(\frac{B'}{B^2}) = 0$. But $B^2 + 4B' \equiv (B_f^2 + 4B_f')^n(B_A^2 + B_A')m \pmod 8$.

Now from the above it is clear that $Tr(\frac{B'}{B^2}) = nTr(\frac{B_f'}{B_f^2}) + mTr(\frac{B_A'}{B_A^2})$.

Now $o = mTr(\frac{B_g'}{B_f^2}) + mTr(\frac{B_A'}{B_A^2})$ iff $nr_f + m\gamma_A \equiv 0 \pmod 2$. Now we can easily see that Theorem 1 is true. Now we have the following

COROLLARY 1

A necessary condition for f(A) to be irreducible is that m be even, and n odd, or that m be odd and r_A be odd.

COROLLARY 2

If p=2 and A(x) is a polynomial as defined for lemma 2, and of degre ≥ 8 then f(A) has an even number of factors and is therefore never irreducible.

PROOF

From corollary 1 we must only check m odd, and see r_A can not be odd. This we can see applying the $GF(2^k)$ version of Stickelberger Theorem to D(A) as computed in Lemma 2.

BIBLIOGRAPHY

1. S. Agou, Factorization sur un corp Fini F_{p^n} des Polynomes Composes $f(x^{p^n} - ax)$, lorsque f(x) est un Polynome Irreductible de $F_{p^n}[X]$, J. Number Theory 9 (1977): 229-239.

2. Irreductibilite des polynomes $f(\sum_{i=0}^{m} a_i X^{p^{ri}})$ sur un corps finit F_{p^s}, Canad. Math. Bull. 23, 207-212 (1980).

3. S. Agou, Irredutibilité des Polynomes $f(x^{p^{2r}} - ax^{p^r} - bx)$ sur un Corps Fini F_{p^s}, J. Number Theory 10 (1978), 64-69.

4. E. R. Berlekamp, Algebraic Coding Theory: New York: McGraw-Hill, 1968.

5. O. Ore, On a Special Class of polynomials, Trans. Am. Math. Soc, 35 (1933), 559-584.

6. O. Ore, Contributions to the Theory of Finite Fields, Trans. Am. Math. Soc., 36 (1934), 243-274.

7. R. G. Swan, Factorization of Polynomials over Finite Fields, Pacific J. Math., 12 (1962), 1099-1106.

COMPUTATIONAL ASPECTS OF REDUCTION STRATEGIES TO CONSTRUCT RESOLUTIONS OF MONOMIAL IDEALS

H.Michael Möller
Fernuniversität
Hagen

Ferdinando Mora
Università
Genova

In (MM1) we showed how the partial knowledge of a resolution of a monomial ideal $M_T(I)$ generated by the maximal terms of a Gröbner basis of a given ideal I in a polynomial ring allows to compute a minimal resolution for the homogenization of I, by means of linear system solutions and Gaussian elimination.

One could choose the Taylor resolution (TAY) which is defined in terms of l.c.m.'s of tuples of generating monomials; if one restricts to pairs, one obtains exactly the "critical pairs" of Buchberger's algorithm (BU1,BU2) for Gröbner basis computation, which in fact is known to give also syzygies.

However the ranks of the Taylor resolution are often very high w.r.t. the actual ranks of the minimal resolution of $M_T(I)$, so that techniques to obtain shorter resolutions without too many computations are of interest.

The well-known Buchberger's criterion (BU3) to avoid the treatment of some critical pairs in his algorithm can be easily interpreted as avoiding to consider some syzygies, which are known to be redundant because a third syzygy can be produced, defining the redundant syzygy in terms of other ones.

This can be generalized to higher modules of syzygies, leading to the concept of "reduction strategy" as a technique to detect redundant generators in higher modules, by producing the generator in the next module which defines it in terms of other generators. So, many unnecessary Gaussian eliminations can be avoided and this is more true since whole sets of redundant generators can be often detected together, so in some stategies there is no need to consider all tuples to detect their redundance.

Once redundant generators are detected, the resulting "reduced" resolution can be obtained by means of a recursive definition.

In this paper we use reduction strategies as a technique to prove the existence of two subresolutions of Taylor's (the second one being an improvement on a similar result of Lyubeznik (LYU)). Then we will present three different reduction strategies and we will discuss briefly the complexity of such algorithms and give heuristical results of their performance on some classes of examples. These heuristics suggest that strateg #2 and strategy #3 are quite effective, giving often resolutions very near to the minimal one, without treating many more tuples. Finally, we will show how to produce a T-resolution for T-hom. modules and an algorithm for Hilbert function based on strategy #3.

ACKNOWLEDGEMENTS We had a correspondence on this subject with David Bayer and Gennady Lyubeznik. Their remarks were helpful for motivating our interest in the problem and for improving some of our results.

1 REDUCTION STRATEGIES

1.1 Let $P:=k[X_1,\dots,X_n]$ a polynomial ring over a field, $I \subset P$ a monomial ideal, $(m_1,\dots,m_r)$ a reduced basis of I consisting of r terms.

Denote:

$I_k := \{(i_1,\dots,i_k) : 1 \leq i_1 < i_2 < \dots < i_k \leq r\}$ for $k=1..r$;

$T(i_1,\dots,i_k) := \text{l.c.m.}(m_{i_1},\dots,m_{i_k})$

$T(i_1,\dots,\not{i}_j,\dots,i_k) := \text{l.c.m.}(m_{i_1},\dots,m_{i_{j-1}},m_{i_{j+1}},\dots,m_{i_k})$

$\tau(i_1,\dots,\not{i}_j,\dots,i_k) := (-1)^j\, T(i_1,\dots,i_k)/T(i_1,\dots,\not{i}_j,\dots,i_k)$

$r_k := \binom{r}{k} = \text{card}(I_k)$.

We can consider the components of P^{r_k} as indexed by elements of I_k; we will write then P^{I_k} for P^{r_k} and denote $e_{(i_1,\dots,i_k)}$ the generic canonical basis element.

Define:

$d_0 : P \to P/I$ the canonical projection

$d_1 : P^{I_1} \to P$ by $d_1(e_i) := m_i$

and, for $2 \leq k \leq r$:

$d_k : P^{I_k} \to P^{I_{k-1}}$ by $d_k(e_{(i_1,\dots,i_k)}) := \sum_j \tau(i_1,\dots,\not{i}_j,\dots i_k)\, e_{(i_1,\dots,\not{i}_j,\dots,i_k)}$

Then:

$$0 \quad P^{I_r} \xrightarrow[d_r]{} P^{I_{r-1}} \to \dots \to P^{I_2} \xrightarrow[d_2]{} P^{I_1} \xrightarrow[d_1]{} P \xrightarrow[d_0]{} P/I \to 0 \quad (1.1)$$

is a free resolution of P/I, the so called Taylor resolution (TAY,DEP,MM1).

If we grade each P^{I_k} by assigning $\deg(e_{(i_1,\dots,i_k)}) := T(i_1,\dots,i_k)$, it is a homogeneous T-resolution of T-graded modules (for terminology cf. (MM1)).

1.2 For k=1..r, let $J_k \subset I_k$ be s.t. for each $(i_1,\dots,i_k) \in J_k$, for each μ, $(i_1,\dots,\not{i}_\mu,\dots,i_k) \in J_{k-1}$, $J_1 = I_1$. Then, for each $k > 1$, if d_k^0 denotes the restriction of d_k to P^{J_k}, $\text{Im}(d_k^0) \subset P^{J_{k-1}}$. If $\text{Ker}\, d_{k-1}^0 = \text{Im}\, d_k^0$, then also:

$$0 \to P^{J_r} \xrightarrow{d_r} P^{J_{r-1}} \to \dots \to P^{J_2} \xrightarrow{d_2} P^{J_1} \xrightarrow{d_1} P \xrightarrow{d_0} P/I \to 0 \quad (1.2)$$

is a free resolution of P/I, a subresolution of Taylor's, which we will call <u>characterized by the sets</u> J_k.

1.3 A <u>reduction strategy</u> for a subresolution (1.2) of Taylor's, characterized by $J_1, \dots, J_r$ (and, in particular, if $J_k = I_k$ for each k, for the Taylor resolution (1.1)) is the assignment of:

1) a partition $J_k =: D_k \cup R_k \cup C_k$ into disjoint subsets for each k=1..r

2) a T-deg compatible ordering on each C_k (i.e. an ordering $\preccurlyeq$ s.t. if $T(i_1,\dots,i_k) < T(j_1,\dots,j_k)$ then $(i_1,\dots,i_k) \preccurlyeq (j_1,\dots,j_k)$

3) a bijection $\Phi_k : C_k \to D_{k+1}$ for each k=2..r-1

such that

R1) $D_1 = D_2 = C_1 = C_r = \emptyset$

R2) if $(i_1,\dots,i_k) \in C_k$, then $T(i_1,\dots,i_k) = T(\Phi_k(i_1,\dots,i_k))$

R3) if $(j_1,\dots,j_{k+1}) = \Phi_k(i_1,\dots,i_k)$, then there is ν: $(i_1,\dots,i_k) = (j_1,\dots,\not{j}_\nu,\dots,j_{k+1})$

R4) if $(j_1,\dots,j_{k+1}) = \Phi_k(j_1,\dots,\not{j}_\nu,\dots,j_{k+1})$ and if for some μ, $(j_1,\dots,\not{j}_\mu,\dots j_{k+1})$ is in C_k and $T(j_1,\dots,\not{j}_\mu,\dots,j_{k+1}) = T(j_1,\dots,j_{k+1})$ then $(j_1,\dots,\not{j}_\mu,\dots,j_{k+1}) \preccurlyeq (j_1,\dots,\not{j}_\nu,\dots,j_{k+1})$.

1.4 Define $p_k : P^{J_k} = P^{D_k} \oplus P^{R_k} \oplus P^{C_k} \to P^{R_k}$ to be the zero morphism on P^{D_k}, the identity on P^{R_k} and on P^{C_k} by:

$$p_k(e_{(i_1,\dots,i_k)}) := (-1)^{\nu-1} \sum_{\mu\neq\nu} (j_1,\dots,\not{j}_\mu,\dots,j_{k+1})\, p_k(e_{(j_1,\dots,\not{j}_\mu,\dots,j_{k+1})})$$

if $(i_1,\dots,i_k) = (j_1,\dots,\not{j}_\nu,\dots,j_{k+1})$ and $(j_1,\dots,j_{k+1}) = \Phi_k(i_1,\dots,i_k)$.

Define $\delta_k : P^{R_k} \to P^{R_{k-1}}$ by $\delta_1 := d_1$ and, if $k \geq 2$:

$$\delta_k(e_{(i_1,\dots,i_k)}) := \sum_\mu \tau(i_1,\dots,\not{i}_\mu,\dots i_k)\, p_{k-1}(e_{(i_1,\dots,\not{i}_\mu,\dots,i_k)})$$

1.5 **LEMMA** Define $a_k : P^{J_k} \to P^{J_k}$ to be the identity on $P^{D_k} \oplus P^{R_k}$ and on P^{C_k} by

$$a_k(e_{(i_1,\dots,i_k)}) := d^0_{k+1}(e_{\Phi_k(i_1,\dots,i_k)})$$

Then:

1) p_k is surjective, a_k an isomorphism

2) $p_k a_k$ is the identity on P^{R_k}, the zero morphism on $P^{C_k} \oplus P^{D_k}$

3) $d'_k := a_{k-1}^{-1} d^0_k a_k$ is zero on P^{C_k}, while if $(j_1,\dots,j_k) \in D_k$:

$$d'_k(e_{(j_1,\dots,j_k)}) = e_{\Phi_{k-1}^{-1}(j_1,\dots,j_k)}$$

4) $\delta_k p_k = p_{k-1} d^0_k$

5) $\mathrm{Im}\, d'_k \cap P^{D_{k-1}} = \emptyset$

6) $d'_k(\mathrm{Ker}\, p_k a_k) = \mathrm{Im}\, d'_k \cap \mathrm{Ker}\, p_{k-1} a_{k-1}$

<u>Proof</u>: 1) if we order the basis of P^{J_k} so that elements indexed in $D_k \cup R_k$ come first, and elements indexed in C_k are ordered according to $\leqslant$, the matrix defining a_k is triangular with ± 1's on the diagonal.

2), 3), 4) require just easy and straightforward verifications

5) if $U = d'_k(V) \in P^{D_{k-1}}$, then $d'_{k-1}(U) = d'_{k-1} d'_k(V) = a^{-1}_{k-1} d^0_{k-1} d^0_k a_k(V) = 0$ implying $U = 0$ since by 3) d'_{k-1} restricted to $P^{D_{k-1}}$ is a bijection onto $P^{C_{k-2}}$.

6) $d'_k(\mathrm{Ker}\, p_k a_k) = d'_k(P^{D_k} \oplus P^{C_k}) = P^{C_{k-1}}$

$\mathrm{Ker}\, p_{k-1} a_{k-1} \cap \mathrm{Im}\, d'_k = (P^{D_{k-1}} \oplus P^{C_{k-1}}) \cap \mathrm{Im}\, d'_k = (P^{D_{k-1}} \cap \mathrm{Im}\, d'_k) \oplus P^{C_{k-1}} = P^{C_{k-1}}$

(the second equality holds by the modular law since $P^{C_{k-1}} \subset \mathrm{Im}\, d'_k$).

1.6 LEMMA In the following commutative diagrams of P-modules

$$\begin{array}{ccccc} M & \xrightarrow{f} & M' & \xrightarrow{g} & M'' \\ \downarrow p & & \downarrow p' & & \downarrow p'' \\ N & \xrightarrow[f']{} & N' & \xrightarrow[g']{} & N'' \end{array}$$

if the upper row is exact, p, p', p'' are surjective and $\mathrm{Ker}\, p'' \cap \mathrm{Im}\, g = g(\mathrm{Ker}\, p')$ then the lower row is exact.

<u>Proof</u>: An immediate consequence of the 3x3 lemma for exact categories (MIT)

1.7 COROLLARY

$$0 \to P^{R_r} \xrightarrow[\delta_r]{} P^{R_{r-1}} \to \ldots \to P^{R_2} \xrightarrow[\delta_2]{} P^{R_1} \xrightarrow[\delta_1]{} P \xrightarrow[\delta_0]{} P/I \to 0 \qquad (1.3)$$

is a free resolution of P/I (namely, δ_k being T-homogeneous, a T-homogeneous resolution). Moreover, if for each $(i_1,\ldots,i_k) \in R_k$, for each μ, $(i_1,\ldots,\not{i}_\mu,\ldots,i_k) \in R_{k-1}$, then it is a subresolution of Taylor's characterized by the sets R_i.

1.8 A particular reduction strategy for the Taylor resolution was presented in (MM1). We then gave the general concept for Taylor resolution only. We owe the suggestion to apply reduction strategies also to subresolution to Lyubeznik.

The notations used here are different from the ones in (MM1), with the aim of making them less cumbersome.

1.9 We will use the following notations (where $J_1,\ldots,J_r$ characterize a subresolution):

$T := \{ \tau : \tau = T(i_1,\ldots,i_k)$ for some $(i_1,\ldots,i_k) \in I_k$, for some $k\}$

$I_{\tau k} := \{(i_1,\dots,i_k) \in I_k : T(i_1,\dots,i_k) = \tau\}$

$J_{\tau k} := I_{\tau k} \cap J_k$; $I_\tau := \bigcup_k I_{\tau k}$; $J_\tau := \bigcup_k J_{\tau k}$

$(i_1,\dots,i_k)\star i$ denotes the tuple $(i_1,\dots,i_j,i,i_{j+1},\dots,i_k)$ with $i_j < i < i_{j+1}$

$\emptyset$ denotes the void tuple so $\emptyset\star i = (i)$

$\deg_j(\tau)$ denotes the degree of τ in the j-th variable

1.10 The obvious way to obtain reduction strategies is simply the following:
Define a T-deg compatible ordering $\leqslant$ on each J_k.
By iteration on k, define $R_k, C_k, D_{k+1}, \Phi_k$ as follows, starting with k=2 and $D_1=C_1=D_2=\emptyset$, $R_1=J_1$:

at any step treat the least $(i_1,\dots,i_k) \in J_k - D_k$ w.r.t. $\leqslant$ which has been not yet treated.

If there is j s.t. $j \neq i_\mu$ for each μ, T(j) divides $T(i_1,\dots,i_k)$ and, denoting $(j_1,\dots,j_{k+1}) := (i_1,\dots,i_k)\star j$, denoting ν the index s.t. $(i_1,\dots,i_k) = (j_1,\dots,\not{j}_\nu,\dots,j_{k+1})$, if for each $(j_1,\dots,\not{j}_\mu,\dots,j_{k+1}) \in J_k - D_k$ s.t. $T(j_1,\dots,\not{j}_\mu,\dots,j_{k+1}) = T(j_1,\dots,j_{k+1})$, $(j_1,\dots,\not{j}_\mu,\dots,j_{k+1}) \leqslant (i_1,\dots,i_k)$ holds, then

assign $(i_1,\dots,i_k)$ to C_k, $(j_1,\dots,j_{k+1})$ to D_{k+1} and define $\Phi_k(i_1,\dots,i_k) := (j_1,\dots,j_{k+1})$

else

assign $(i_1,\dots,i_k)$ to R_k.

Its main drawback is that all tuples are to be treated by this procedure, while it is possible (as it will be done in the strategies presented here) to assign whole sets of tuples to C_k and D_k without actually constructing them.

2 SUBRESOLUTIONS OF TAYLOR'S

2.1 We call $\tau \in T$ an inner term if there is i s.t. T(i) divides τ and $\deg_j(T(i)) < \deg_j(\tau)$ if $\deg_j(\tau) > 0$; it is called outer otherwise. If τ is inner denote $i(\tau)$ such an i verifying the above condition. Denote OT the subset of T consisting of outer terms, IT the one consisting of inner terms.
Also, we say $(i_1,\dots,i_k)$ is contained in $(j_1,\dots,j_l)$ if $\{i_1,\dots,i_k\} \subset \{j_1,\dots,j_l\}$.

2.2 For each k, denote $M_k := \bigcup_{\tau\in OT} I_{\tau k}$, $L_k := \{(i_1,\dots,i_k) \in I_k$ s.t. for all $t \leqslant k$, for all $q > i_t$, T(q) doesn't divide $T(i_1,\dots,i_t)\}$, $N_k := L_k \cap M_k$.
The aim of this paragraph is to show that the sets M_k, L_k, N_k characterize subresolutions of Taylor's.

2.3 PROPOSITION $M_1,\dots,M_r$ characterize a subresolution of Taylor's

Proof: If τ is an inner term, I_τ can be partitioned $I_\tau = C_\tau \cup D_\tau$, where

$C_\tau := \{\alpha : \alpha \in I_\tau$, not containing $i(\tau)\}$, $D_\tau := \{\alpha \in I_\tau$, containing $i(\tau)\}$.

Let $C_{\tau k} := C_\tau \cap I_k$, $D_{\tau k} := D_\tau \cap I_k$. There are then bijections $\Phi_{\tau k}: C_{\tau k} \to D_{\tau k+1}$, defined by $\Phi_{\tau k}(\alpha) := \alpha \star i(\tau)$.

Then, if $C_k := \bigcup_\tau C_{\tau k}$, $D_k := \bigcup_\tau D_{\tau k}$, $I_k = D_k \cup M_k \cup C_k$ is a partition, $\Phi_k: C_k \to D_{k+1}$ defined by $\Phi_k(\alpha) := \Phi_{\tau k}(\alpha)$ if $\tau = T(\alpha)$ is a bijection.

Moreover if $(j_1,\dots,j_{k+1}) = \Phi_k(j_1,\dots,\not{j}_\nu,\dots,j_{k+1})$, then $\tau := T(j_1,\dots,j_{k+1}) \in IT$, so $j_\nu = i(\tau)$ and therefore if $T(j_1,\dots,\not{j}_\mu,\dots,j_{k+1}) = \tau$, with $\mu \neq \nu$, $(j_1,\dots,\not{j}_\mu,\dots,j_{k+1})$ is in D_k, since it contains $i(\tau)$.

Therefore, if $\preccurlyeq$ is any T-deg compatible ordering on C_k, R1-4) hold trivially.

Moreover, if $\alpha := (i_1,\dots,i_k) \in M_k$, $\beta := (i_1,\dots,\not{i}_\mu,\dots,i_k)$, then $T(\beta)$ divides $T(\alpha) \in \in OT$, so also $T(\beta) \in OT$, $\beta \in M_{k-1}$.

This proves $M_1,\dots,M_r$ characterize a subresolution of Taylor's because of cor.1.7.

2.4 PROPOSITION (Lyubeznik: (LYU)) $L_1,\dots,L_r$ characterize a subresolution of Taylor's.

2.5 PROPOSITION $N_1,\dots,N_r$ characterize a subresolution of Taylor's.

Proof: We will give partitions $I_{\tau k} = D_{\tau k} \cup N_{\tau k} \cup C_{\tau k}$, bijections $\Phi_{\tau k}: C_{\tau k} \to D_{\tau k+1}$, orderings $\preccurlyeq_\tau$ on each $C_{\tau k}$.

Then, defining $C_k := \bigcup_\tau C_{\tau k}$, $D_k := \bigcup_\tau D_{\tau k}$, $\Phi_k: C_k \to D_{k+1}$ by $\Phi_k(\alpha) := \Phi_{\tau k}(\alpha)$ if $\tau = T(\alpha)$, and an ordering $\preccurlyeq$ on each C_k by $\alpha \preccurlyeq \beta$ if $T(\alpha) < T(\beta)$ or $T(\alpha) = T(\beta) =: \tau$ and $\alpha \preccurlyeq_\tau \beta$, we will show R1-4).hold. Since $(i_1,\dots,i_k) \in N_k$ implies $(i_1,\dots,\not{i}_\mu,\dots,i_k) \in N_{k-1}$ (N_k is the intersection of L_k and M_k, where this property is verified), this will prove the thesis.

If $\tau \in IT$, define $C_{\tau k}$, $D_{\tau k}$, $\Phi_{\tau k}$ as in 2.1, let $\preccurlyeq_\tau$ be any ordering.

If $\tau \in OT$, let $\alpha := (i_1,\dots,i_k) \in I_\tau - N_\tau$.

Let t be the highest index s.t. there is $q' > i_t$ s.t. $T(q')$ divides $T(i_1,\dots,i_t)$ and let q be the highest such q'. If $t < k$ and $q = i_{t+1}$, then assign α to $D_{\tau k}$, if $t = k$ or $q < i_{t+1}$, then assign α to $C_{\tau k}$ and define $\Phi_{\tau k}(\alpha) := \alpha \star q \in I_\tau - N_\tau$.

If $\beta \in D_{\tau k+1}$, $\beta := (i_1,\dots,i_{k+1})$, there is t s.t. $T(i_1,\dots,i_t) = T(i_1,\dots,i_{t+1})$ and if $q > i_s$, with $s > t$, then $T(q)$ doesn't divide $T(i_1,\dots,i_s)$. This implies $\alpha := := (i_1,\dots,\not{i}_{t+1},\dots,i_{k+1})$ is in $C_{\tau k}$ and $\Phi_{\tau k}(\alpha) = \beta$. So $\Phi_{\tau k}$ is surjective.

If $(i_1,\dots,i_{k+1}) = \Phi_{\tau k}(i_1,\dots,\not{i}_\mu,\dots,i_{k+1}) = \Phi_{\tau k}(i_1,\dots,\not{i}_\nu,\dots,i_{k+1})$, $\mu < \nu$, then $T(i_1,\dots,i_{\mu-1}) = T(i_1,\dots,i_\mu)$, $T(i_1,\dots,i_{\nu-1}) = T(i_1,\dots,i_\nu)$, which implies $T(i_1,\dots,\not{i}_\mu,\dots,i_{\nu-1}) = T(i_1,\dots,i_{\nu-1}) = T(i_1,\dots,i_\nu) = T(i_1,\dots,\not{i}_\mu,\dots i_\nu)$ and μ doesn't satisfy the required maximality conditions giving a contradiction and proving $\Phi_{\tau k}$ is bijective.

If $\tau \in OT$, $\alpha,\beta \in C_{\tau k}$, $\alpha := (i_1,\dots,i_k)$, $\beta := (j_1,\dots,j_k)$, then let t be the least index s.t. if $s > t$, $T(i_1,\dots,i_s) = T(j_1,\dots,j_s)$ and $i_s = j_s$. Then define $\alpha \leqslant_\tau \beta$ iff $T(i_1,\dots,i_t) < T(j_1,\dots,j_t)$ or $T(i_1,\dots,i_t) = T(j_1,\dots,j_t)$ and $i_t > j_t$.

Let us then show that R4 holds (R1-3) hold trivially). So, let $\alpha,\beta \in C_k$, $(j_1,\dots,j_{k+1}) := := \Phi_k(\alpha)$, $\alpha = (j_1,\dots,\not{j}_\nu,\dots,j_{k+1}) =: (i_1,\dots,i_k)$, $\beta = (j_1,\dots,\not{j}_\mu,\dots,j_{k+1}) =: =: (i'_1,\dots,i'_k)$, $\tau := T(\alpha) = T(\beta)$. We want to show $\beta \leqslant \alpha$.

If $\tau \in IT$ see the proof of 2.3.

If $\tau \in OT$, $\nu \quad \mu$ would be in contradiction with $\Phi_k(\alpha) = (j_1,\dots,j_{k+1})$ since then $T(j_1,\dots,\not{j}_\nu,\dots j_{\mu-1}) = T(j_1,\dots,\not{j}_\nu,\dots,j_\mu)$. So $\mu < \nu$. Then if $\lambda > \nu$, $T(j_1,\dots,\not{j}_\nu,\dots,j_\lambda) = = T(j_1,\dots,\not{j}_\mu,\dots,j_\lambda)$. Also $T(j_1,\dots,\not{j}_\mu,\dots,j_\nu) = T(j_1,\dots,j_{\nu-1})$ and $j_\nu > j_{\nu-1}$, implying $\beta \leqslant \alpha$.

3 REDUCTION ALGORITHMS

In this paragraph, we present three different algorithms, based on different reduction strategies, to compute resolutions (1.3). In view of the recursive definitions (cf.1.4) of p_k and δ_k, we observe explicitly that:

1) if s denotes the first index s.t. $R_s = \emptyset$, the construction requires just the explicit knowledge of Φ_k and p_k for $k \leqslant s-2$ and of δ_k for $k \leqslant s-1$. To find such an s, it is necessary to know also Φ_{s-1} (to obtain D_s) and Φ_s (to obtain C_s).

2) there is also no need to build explicitly $\leqslant$, D_k, C_k and Φ_k. All is needed is:

   a) a procedure to list each R_k up to s

   b) a procedure which, for any given tuple, assigns it to the subset to which it belongs

   c) a procedure that, for any given tuple in C_k, produces its image under Φ_k.

So for each strategy discussed here, we will first present such procedures and then define an ordering $\leqslant$ to show that conditions R1-4) are verified.

The first strategy is modelled after the second algorithm for Hilbert function computat we presented in (MM2). The second one (whose author is Möller and which is described in detail in (MOL)) is modelled after the third algorithm there; it works separately on each I_τ, so it can be restricted only to outer terms, thus giving a reduction for the subresolution characterized by M_k. The third one applies the direct approach outlined in 1.10 to the subresolution characterized by N_k and consists mainly in an efficient construction of these sets.

3.1 STRATEGY 1 (iteration on generators)

3.1.1 We denote $I_{tk} := \{(i_1,\dots,i_k) \in I_k : i_k < t\}$ for $t=2..r+1$. The following algorithm constructs subsets $R_k(t)$ of I_{tk} for each t. Finally we will pose $R_k := R_k(r+1)$.

Define $R_1(2) := \{1\}$, $R_k(2) := \emptyset$ for $k \geqslant 2$.

Then, by iteration on t, construct $R_k(t+1)$ as follows:

Let $J_k(t) := \{(i_1,\dots,i_{k-1},t) : (i_1,\dots,i_{k-1}) \in R_{k-1}(t)\}$ and impose a T-deg compatible ordering $\langle$ on each $J_k(t)$.

By iteration on k construct partitions $J_k(t) =: D_k^\star \cup R_k^\star \cup C_k^\star$, $R_k(t) =: S_k \cup B_k$ and bijections $\psi_{kt}: C_k^\star \cup B_k \to D_k^\star$ as follows:

1) if $T(i_1,\dots,i_k,t) = T(i_1,\dots,i_k)$ assign $(i_1,\dots,i_k)$ to B_k, $(i_1,\dots,i_k,t)$ to $D_k^\star$ and let $\psi_{kt}(i_1,\dots,i_k) = (i_1,\dots,i_k,t)$

2) otherwise, if $(i_1,\dots,i_k,t)$ is s.t.

i) if $T(i_1,\dots,\not{i}_\nu,\dots,i_k,t) = T(i_1,\dots,i_k,t)$ and $(i_1,\dots,\not{i}_\nu,\dots,i_k)$ is not in $R_k(t)$, then $(i_1,\dots,\not{i}_\nu,\dots,i_k) \in D_{k-1}$ (this can be verified by 3.1.2 below)

ii) there is μ s.t. $T(i_1,\dots,\not{i}_\mu,\dots,i_k,t) = T(i_1,\dots,i_k,t)$ and $(i_1,\dots,\not{i}_\mu,\dots,i_k,t) = \max_\langle \{(i_1,\dots,\not{i}_\nu,\dots,i_k,t) \in J_k(t)-D_k\}$

discard $(i_1,\dots,\not{i}_\mu,\dots,i_k,t)$ from R_k and assign it to C_k, assign $(i_1,\dots,i_k)$ to S_k, $(i_1,\dots,i_k,t)$ to $D_k^\star$, let $\psi_{kt}(i_1,\dots,\not{i}_\mu,\dots,i_k,t) := (i_1,\dots,i_k,t)$

3) otherwise, assign $(i_1,\dots,i_k)$ to S_k, $(i_1,\dots,i_k,t)$ to $R_k^\star$.

Let $R_k(t+1) := R_k^\star \cup S_k$.

3.1.2 If $(i_1,\dots,i_k) \notin R_k$, then let μ be the highest index s.t. $(i_1,\dots,i_\mu)$ belongs to some $R_\mu(u)$ and t be the highest such u. Then if:

1) $t < i_{\mu+1}$: $(i_1,\dots,i_k) \in C_k$ and $\Phi_k(i_1,\dots,i_k) = (i_1,\dots,i_k)\star t$

2) $t = i_{\mu+1}$ and $(i_1,\dots,i_{\mu+1}) \in D_{\mu+1}^\star$: $(i_1,\dots,i_k) \in D_k$

3) $t = i_{\mu+1}$ and $(i_1,\dots,i_{\mu+1}) \in C_{\mu+1}^\star$: $(i_1,\dots,i_k) \in C_k$ and $\Phi_k(i_1,\dots,i_k) :=$
$:= (i_1,\dots,i_k)\star u$ if $\psi_{\mu+1t}(i_1,\dots,i_{\mu+1}) = (i_1,\dots,i_{\mu+1})\star u$.

3.1.3 Denote $C_k(t) := \{(i_1,\dots,i_k) \in C_k : i_k < t\}$, $Q_k(t) := \{(i_1,\dots,i_k,t) : (i_1,\dots,i_k) \in C_k(t)\}$ and remark that $C_k(t+1) = C_k^\star \cup B_k \cup C_k(t) \cup Q_k(t)$.

Define inductively an ordering $\leqslant_t$ on each $C_k(t+1)$ as follows:

elements in $C_k^\star$ are first, followed orderly by elements in $Q_k(t)$, B_k, $C_k(t)$; while, in each subset, $\leqslant_t$ is defined to be $\leqslant_{t-1}$ in $C_k(t)$, the order naturally induced by it in $Q_k(t)$, $\langle$ in $C_k^\star$, while any ordering can be used in B_k.

Finally let $\leqslant := \leqslant_r$.

3.1.4 LEMMA R1-4 hold under these definitions

Proof: We have just to prove R4. So let $(i_1,\dots,i_k) \in C_k$ and let t be s.t. $(i_1,\dots,i_k)$ is in $C_k(t+1) - C_k(t)$. Then:

i) if $(i_1,\dots,i_k) \in C_k^\star$, then R4 holds by construction

ii) if $(i_1,\dots,i_k) \in Q_k(t)$, then $i_k = t$. So, let $(j_1,\dots,j_k) := \Phi_{k-1}(i_1,\dots,i_{k-1})$; then $\Phi_k(i_1,\dots,i_k) = (j_1,\dots,j_k,t)$. Therefore $(j_1,\dots,j_k) \in D_k$ and if $(j_1,\dots,\not{j}_\nu,\dots,j_k,t) \in C_k - C_k^\star$, it belongs to $Q_k(t)$. Inductively we may assume $(j_1,\dots,\not{j}_\nu,\dots,j_k) \preccurlyeq (i_1,\dots,i_{k-1})$, implying $(j_1,\dots,\not{j}_\nu,\dots,j_k,t) \preccurlyeq (i_1,\dots,i_k)$.

iii) if $(i_1,\dots,i_k) \in B_k$, $\Phi_k(i_1,\dots,i_k) = (i_1,\dots,i_k,t)$ and if $(i_1,\dots,\not{i}_\nu,\dots,i_k,t)$ is in C_k, then it belongs either to $C_k^\star$ or to $Q_k(t)$.

3.2 STRATEGY #2

3.2.1 This algorithm is devised to partition each M_τ into disjoint subsets D_τ, R_τ, C_τ. Then, denoting $D_{\tau k} := D_\tau \cap M_k$, $R_{\tau k} := R_\tau \cap M_k$, $C_{\tau k} := C_\tau \cap M_k$, D_k is obtained as $D_k := \bigcup_\tau D_{\tau k}$, R_k and C_k are obtained analogously.

3.2.2 If τ is outer, by means of a recursive procedure NEXTINDEX described below, a sequence of tuples $\Lambda := (\alpha_1,\dots,\alpha_l)$ is produced s.t., denoting $A_\nu := \{\alpha \in I_\tau : \alpha$ contains α_ν and dosn't contain α_μ if $\mu < \nu\}$, then $I_\tau = \bigcup_\nu A_\nu$ and either $A_\nu = \{\alpha_\nu\}$ or A_ν is s.t. $\alpha_\nu \in A_\nu$, and there is $i(\tau,\nu)$ s.t. $\beta \in A_\nu$ and doesn't contain $i(\tau,\nu)$ iff $\beta\star i(\tau,\nu) \in A_\nu$. Also, the procedure allows to know which of the two cases occurs and returns such an $i(\tau,\nu)$.

3.2.3 Procedure NEXTINDEX returns such a sequence $\Lambda := (\alpha_1,\dots,\alpha_l)$ and for each ν it conveys the information which of the two possibilities for A_ν occurs. Also, if the second one occurs, $i(\tau,\nu)$ is returned.

It depends on four parameters: a subset J of the variable indexes, a subset K of the basis element indexes, a sequence X of tuples and a tuple α.

It is called with initial parameters $J := \{j : \deg_j(\tau) > 0\}$, $K := \{k : T(k)$ divides $\tau\}$, X the void sequence, α the void tuple.

```
NEXTINDEX(J,K,X,α)
K' := K
let j ∈ J
while K_1 := {k ∈ K' : deg_j(T(k)) = deg_j(τ)} ≠ Ø do
      let k ∈ K_1
      K' := K' - {k}
      X := X ∪ {α★k}
```

$J' := \{i \in J : \deg_i(T(k)) = \deg_i(\tau)\}$

If $K' \neq \emptyset$ and $J - J' \neq \emptyset$ then call NEXTINDEX$(J-J', K', X, \alpha \star k)$

At termination, Λ is obtained as the subsequence of X consisting of those tuples in M_τ; these are those tuples for which $J-J' = \emptyset$ and, for any such tuple α_ν, $\mathrm{card}(A_\nu) = 1$ iff $K' = \emptyset$, $\mathrm{card}(A_\nu) > 1$ iff $K' \neq \emptyset$, in which case any element in K' can be chosen as $i(\tau,\nu)$.

3.2.4 For each outer term τ, R_τ is defined to consist of those α_ν s.t. $\mathrm{card}(A_\nu) = 1$. If $\beta \in M_k - R_k$, then there is ν s.t. α_ν is contained in β, α_μ is not contained in it if $\mu < \nu$. Then if $i(\tau,\nu)$ is contained in β, $\beta \in D_k$, otherwise $\beta \in C_k$, $\Phi_k(\beta) = \beta \star i(\tau,\nu)$.

3.2.5 We omit proofs which can be found in (MOL), where a detailed presentation of this strategy is given.

3.3 STRATEGY #3

3.3.1 LEMMA If $\tau \in OT$, let $j(\tau)$ be the highest index j s.t. $T(j)$ divides τ.
Let $\Sigma := \{\sigma \in T : \sigma \neq \tau,\ \mathrm{l.c.m.}(\sigma, T(j(\tau))) = \tau\}$.
Then $N_\tau = \bigcup_{\sigma \in \Sigma} \{\alpha \star j(\tau) : \alpha \in N_\sigma\}$.

<u>Proof</u>: if $(i_1,\ldots,i_k) \in N_\sigma$, for a $\sigma \in \Sigma$, then $(i_1,\ldots,i_{k+1}) \in M_\tau$, with $i_{k+1} := j(\tau)$. So suppose it is not in L_τ. Then there is $t \leq k+1$ and $q > i_t$ s.t. $T(i_1,\ldots,i_t) = T(i_1,\ldots,i_t,q)$. Since $j(\tau)$ is the highest index q' s.t. $T(q')$ divides τ, then $q < i_{k+1}$, $t \leq k$, implying $(i_1,\ldots,i_k) \notin L_k$, against the assumption that $(i_1,\ldots,i_{k+1})$ is in L_{k+1}. So $T(i_1,\ldots,i_k) =: \sigma$ divides τ and $\mathrm{l.c.m.}(\sigma, T(j(\tau))) = \tau$

3.3.2 Then the following procedure produces the sets N_τ, $\tau \in OT$.
Fix a term ordering on OT.
Then, for each $\tau \in OT$, in increasing order:

i) if there is a (unique) i s.t. $T(i) = \tau$ then $N_\tau := \{(i)\}$.

ii) otherwise let $\Sigma := \{\sigma \in OT : \mathrm{l.c.m.}(\sigma, j(\tau)) = \tau\}$. Then $N_\tau := \bigcup_\Sigma \{\alpha \star j(\tau) : \alpha \in N_\sigma\}$

Then apply the procedure outlined in 1.10 to obtain a reduction strategy for N_k.

4 SOME REMARKS ON COMPLEXITY

4.1 If we take as measure of the efficiency of the algorithms, the length and ranks of the resulting resolutions and the number of tuples generated during their performance, the results are disappointing for all strategies, except #2, since it is impossible to improve on the obvious bound given by the length and ranks of the Taylor resolution, a nd one can produce examples in which these bounds are nearly reached.

In contrast, since in strategy #2 procedure NEXTINDEX works in such a way that any time the length of the tuple under construction is increased by 1, then the cardinality of J (bounded by n) decreases by 1 at least, one can conclude:
The length of any resolution obtained by strategy #2 is at most n, and no tuples with more than n elements are produced during the algorithm's performance

4.2 The main application of the algorithms is the computation of a minimal resolution either of a monomial ideal or of a homogeneous ideal of which a Gröbner basis is known (for the second question, Bayer (BAY) proposed a different algorithm, which can advance serious claims to be more efficient).
In view of such applications, not a whole resolution must be obtained but just an initial segment of it, whose length exceeds by 1 the length of the minimal resolution. Because of this, and since all strategies can be so applied to produce first all tuples in R_1, then in R_2, and so on, the length of the whole resolution appears of less relevance than the ranks of the first n+1 modules and the cardinality of the generated k-tuples, with $k \leq n+1$.
In this respect, there is no hope to improve on the obvious bounds $\binom{r}{k}$ for any of the strategies (such a bound can be actually reached by the ranks of minimal resolutions if $r \leq n$: take the ideal $(X_1,\dots,X_r)$)

4.3 The proposed efficiency measure doesn't take into account the computations to decide the subset to which a given tuple belongs, the image of such a tuple under Φ_k, the recursive computations required by the definition of p_k.

4.4 All strategies depend on some choices which can strongly affect their performance. All of them (#2 excepted) depend on the ordering of the basis elements and on some ordering of tuples. There seems to be no general rule to give optimal choices.
Strategy #2 depends just on the choice of the variables in the recursion steps of NEXTINDEX. A good heuristical choice is to select a variable index j in J for which the cardinality of K_1 is minimal.

5 HEURISTICS ON THE ALGORITHMS' PERFORMANCE

5.1 In this paragraph we present, for several classes of examples, values for the length of the resolution, cardinality of R_k, with $k \leq n+1$, cardinality of the sets of k-tuples generated with $k \leq n+1$ (denoted respectively l, r_k, s_k) for all strategies presented here, compared with length and ranks of the minimal resolution.
When possible, with values are given as functions of the variable parametrizing the

class of ideals; otherwise the values corresponding to the first values of the parameter are given.
Basis elements are ordered by the graduated-lexicographical term ordering (unless otherwise specified); tuples by the order in which they are generated; if different choices are possible for $\Phi_k(\alpha)$, the highest tuple is chosen.

5.2 These heuristics support the claim that strategy #2 is as good in practice as in theory, that #3 can compete with it in most cases (however, there is always the risk of its exponential behaviour) and that both of them tend to give resolutions which are minimal or very near to minimal. The efficiency of strategy #2 is greater in high-degree-few-variable cases, less in low-degree-many-variable cases where strategy #3 is the most efficient.

5.3 The ideal generated by deg d terms in 2 variables

	1	r_1	r_2	s_1	s_2	s_3
MIN.	2	d+1	d			
1	2	d+1	d	d+1	$\binom{d+1}{2}$	$\binom{d}{2}$
2	2	d+1	d	d+1	d	
3	2	d+1	d	d+1	d	

5.4 The ideal generated by deg d terms in 3 variables

	d	1	r_1	r_2	r_3	r_4	s_1	s_2	s_3	s_4	s_5
MIN		3	$\binom{d+2}{2}$	d^2+2d	$\binom{d+1}{2}$						
1	2	4		8	4	1		15	14	4	
1	3	6	$\binom{d+2}{2}$	15	11	9	$\binom{d+2}{2}$	45	55	27	11
1	4	?		24	27	40		105	146	106	97
2		3	$\binom{d+2}{2}$	d^2+2d	$\binom{d+1}{2}$		$\binom{d+2}{2}$	$\frac{9}{2}d^2-\frac{9}{2}d+3$	$\frac{3}{2}d^2+\frac{3}{2}d-3$		
3		3	$\binom{d+2}{2}$	d^2+2d	$\binom{d+1}{2}$		$\binom{d+2}{2}$	$\frac{5}{2}d^2-\frac{d}{2}+2$	$2d^2-2d+2$		

5.5 The ideal generated by deg 2 squarefree terms in n variables (lexicographically ordered)

	1	r_1	r_2	r_3	s_1	s_2	s_3	s_4
n=4								
MIN	3	6	8	3				
1	3	6	8	3	6	15	13	3
2	3	6	8	3	6	15	10	
3	3	6	8	3	6	9	4	

	1	r_1	r_2	r_3	r_4	r_5	s_1	s_2	s_3	s_4	s_5	s_6
n=5												
MIN	4	10	20	15	4							
1	4	10	20	16	5		10	45	64	34	6	
2	4	10	20	15	4		10	45	81	41		
3	4	10	20	16	5		10	25	24	8		
n=6												
MIN	5	15	40	45	24	5						
1	5	15	40	51	34	9	15	105	204	175	70	18
2	5	15	40	45	24	5	15	105	250	496	225	
3	5	15	40	51	34	9	15	55	85	60	16	

5.6 The ideal $M_{T_2}(J)$, where J is the ideal of the curve $(t^{d^2}, t^{d^2-d+1}, t^{d^2+1})$, i.e. the ideal $(X^{d^2+1}, \{X^{d^2-id+1}Y^i : i=1..d-1\}, Y^{d+1}, YZ^{d-1})$

	1	r_1	r_2	r_3	s_1	s_2	s_3
MIN	3	d+3	2d+2	d			
1	3	d+3	2d+2	d	d+3	$\binom{d+3}{2}$	$\binom{d+2}{2}$
2	3	d+3	2d+2	d	d+3	2d+3	d+1
3	3	d+3	2d+2	d	d+3	2d+3	d+1

5.7 Reisner's example (char. free)

	1	r_1	r_2	r_3	r_4	s_1	s_2	s_3	s_4	s_5
MIN	4	10	15	7	1					
1	4	10	15	9	3	10	45	55	25	6
2	4	10	16	8	1	10	45	75	17	
3	4	10	15	8	2	10	27	27	9	

6 COMPUTING RESOLUTIONS OF MODULES

6.1 Let U be a submodule of a free finite rank module over a polynomial ring. Let $<$ be a term ordering and $V := M_T(U)$. If a T-homogeneous resolution of V is known, the techniques of (MM1) can be applied to obtain minimal T- and H-resolutions of U and a minimal homogeneous resolution of U if U is homogeneous.
We will show in this paragraph how to compute such a resolution for V, using the techniques of this paper.

6.2 Let $P := Q := K[X_1,\dots,X_n]$, $U \subset P^r = P^{r-1} \oplus Q$. Terms of P^r are elements $(m_1,\dots,m_r)$, where there is i s.t. m_i is a term of P, $m_j = 0$ if $j \neq i$, and are in obvious biunivocal correspondence with pairs (m,i) where m is a term of P and $i \in \{1,\dots,r\}$.
Impose a term ordering $\leqslant$ on P^r as follows: $(m,i) \leqslant (n,j)$ iff $m < n$ or $m = n$ and $i < j$.
If $(F_1,\dots,F_t)$ is a G-basis of U w.r.t. $\leqslant$, then $G := (M_T(F_1),\dots,M_T(F_t))$ is a G-basis of V w.r.t. $\leqslant$. Moreover $G \cap P^{r-1}$ is a G-basis of the T-homogeneous submodule of P^{r-1}

$V \cap P^{r-1} =: V_1$, and consists of elements $(f_1,\dots,f_s)$ whose last coordinate is zero, while the remaining elements of G, $g_1,\dots,g_u$, can be written $g_i =: h_i + m_i$, with $h_i \in P^{r-1}$, $0 \neq m_i = \mathrm{Hterm}(g_i) \in Q$.
So $(m_1,\dots,m_u)$ is canonically isomorphic to a T-hom. ideal of Q, and a T-hom. resolution of it can be obtained by reduction strategy techniques.

6.3 Assume then we are given:

i) a T-hom. resolution of V_1 (obtained iteratively by the techniques we are going to describe):

$$\dots \underset{d_3}{\to} P^{t_2} \underset{d_2}{\to} P^{t_1} \underset{d_1}{\to} P^{r-1} \underset{d_0}{\to} P^{r-1}/V \to 0$$

ii) a T-hom. resolution of I (by reduction strategies):

$$\dots \underset{\delta_3}{\to} Q^{s_2} \underset{\delta_2}{\to} Q^{s_1} \underset{\delta_1}{\to} Q \underset{\delta_0}{\to} Q/I \to 0$$

We will show how to construct a T-hom. resolution of V:

$$\dots \underset{D_3}{\to} P^{t_2} \oplus Q^{s_2} \underset{D_2}{\to} P^{t_1} \oplus Q^{s_1} \underset{D_1}{\to} P^{r-1} \oplus Q \to P^r/V \to 0$$

We will denote $(e_{1j},\dots,e_{t_j j})$ a canonical basis of P^{t_j}, $(E_{1j},\dots,E_{s_j j})$ a canonical basis of Q^{s_j}.

6.4 Define $D_1(e_{i1}) := d_1(e_{i1}) = f_i$; $D_1(E_{i1}) := g_i$.
Then the following hold:

1) $(D_1 - \delta_1)(Q) \subset P^{r-1}$
2) D_1 restricted to P^{r-1} is d_1
3) $D_0 D_1 = 0$
4) $\mathrm{Ker}\, D_0 \subset \mathrm{Im}\, D_1$
5) D_1 is T-homogeneous
6) $D_1\delta_2(E_{i2}) \in \mathrm{Im}\, d_1$ for each i

Proof of 6): $D_1\delta_2(E_{i2}) = \delta_1\delta_2(E_{i2}) + (D_1 - \delta_1)\delta_2(E_{i2}) \in P^{r-1}$
So $D_1\delta_2(E_{i2}) \in \mathrm{Im}\, D_1 \cap P^{r-1} = \mathrm{Ker}\, D_0 \cap P^{r-1} = \mathrm{Ker}\, d_0 = \mathrm{Im}\, d_1$.

6.5 Assume iteratively $D_1,\dots,D_i$ have been defined so that:

1) $(D_i - \delta_i)(Q^{s_i}) \subset P^{t_{i-1}}$
2) D_i restricted to P^{t_i} is d_i
3) $D_{i-1}D_i = 0$
4) $\mathrm{Ker}\, D_{i-1} \subset \mathrm{Im}\, D_i$
5) D_i is T-hom.
6) $D_i\delta_{i+1}(E_{ji+1}) \in \mathrm{Im}\, d_i$.

Define then: $D_{i+1}(e_{ji+1}) := d_{i+1}(e_{ji+1})$
Let $F_j := D_i\delta_{i+1}(E_{ji+1}) \in \mathrm{Im}\, d_i$ and let $F_j := \Sigma\, h_l\, d_i(e_{li})$ a T-hom. representation.
Define then $D_{i+1}(E_{ji+1}) := \delta_{i+1}(E_{ji+1}) - \Sigma\, h_l e_{li}$.
Then the following hold:

1) $(D_{i+1} - \delta_{i+1})(E_{ji+1}) \in P^{t_i}$

2) D_{i+1} restricted to $P^{t_{i+1}}$ is d_{i+1}

3) $D_i D_{i+1} = 0$

4) $\mathrm{Ker}\, D_i \subset \mathrm{Im}\, D_{i+1}$

5) D_{i+1} is T-homogeneous

6) $D_{i+1\, i+2}(E_{ji+2}) \in \mathrm{Im}\, d_{i+1}$

Proof: 4) Let $F \in \mathrm{Ker}\, D_i$, $F = \Sigma h'_l e_{li} + \Sigma h''_k E_{ki}$.

So $0 = D_i(F) = (\Sigma h'_l d_i(e_{li}) + \Sigma h''_k (D_i - \delta_i)(E_{ki})) + \Sigma h''_k \delta_i(E_{ki})$, the first summand being in $P^{t_{i-1}}$, the second in $\Omega^{s_{i-1}}$.

So $\Sigma h''_k \delta_i(E_{ki}) = 0$, $\Sigma h''_k E_{ki} \in \mathrm{Ker}\, \delta_i = \mathrm{Im}\, \delta_{i+1}$; therefore $F = \Sigma h'_l e_{li} + \delta_{i+1}(G_1)$ for some G_1 in $\Omega^{s_{i+1}}$.

So $G_2 := G - D_{i+1}(G_1) = \Sigma h'_l e_{li} - (D_{i+1} - \delta_{i+1})(G_1) \in P^{t_i}$ and $0 = D_i(F) = D_i(F - D_{i+1}(G_1)) =$ $= D_i(G_2) = d_i(G_2)$; therefore $G_2 = d_{i+1}(G_3) = D_{i+1}(G_3)$ for some G_3 and $F =$ $= D_{i+1}(G_1) + D_{i+1}(G_3)$ as claimed.

6) $D_{i+1}\delta_{i+2}(E_{ji+2}) = (D_{i+1} - \delta_{i+1})\delta_{i+2}(E_{ji+2}) + \delta_{i+1}\delta_{i+2}(E_{ji+2}) = (D_{i+1} - \delta_{i+1})\delta_{i+2}(E_{ij+2})$ is in $P^{t_i} \cap \mathrm{Im}\, D_{i+1} = P^{t_i} \cap \mathrm{Ker}\, D_i = \mathrm{Ker}\, d_i = \mathrm{Im}\, d_{i+1}$.

7 A NEW ALGORITHM FOR HILBERT FUNCTION COMPUTATION

Her we present an algorithm for Hilbert function computations, adopting freely notations and terminology of (MM2), which is based on the subresolution of Taylor's characterized by the sets N_k and on strategy #3. Essentially it makes use of the formula (easy to be proved using Lemma 3.3.1):

$$\mu_\tau := \sum_k (-1)^k \operatorname{card}(N_{\tau k}) = \sum_{\sigma/\tau} -\mu_\sigma$$

Its complexity is asyntotically $\operatorname{card}(T)^2$ and since $\operatorname{card}(T) \leq r^{n+1}$, r the cardinality of the basis, the asyntotical complexity is at most r^{2n+2}.

The algorithm requires a reduced basis for the monomial ideal as input:

ALGORITHM

```
S := {(a_0,...,a_n) : a_i = deg_i(T(j)) for a j : 1 ≤ j ≤ r}
μ_1 := 1
T := {1}
while S ≠ ∅ do
      let (a_0,...,a_n) := minS (with respect to any term ordering)
      S := S - {(a_0,...,a_n)}
      τ := X_0^a_0 ...X_n^a_n
      if {i : T(i) divides τ} ≠ ∅ then
            j:= max {i : T(i) divides τ}
            μ_τ := 0
            for each σ ∈ T do
```

if l.c.m.$(\sigma, T(j)) = \tau$ then $\mu_\tau := \mu_\tau - \mu_\sigma$

if $\mu_\tau \neq 0$ then $T := T \cup \{\tau\}$

REFERENCES

(BAY) D.BAYER The division algorithm and the Hilbert scheme, Ph.D. Thesis, Harvard 1982

(BU1) B.BUCHBERGER Ein Algorithmus zu auffinden der Basiselemente des Restlassenringe nach einem nulldimensionalen Polynomideal, Ph.D. Thesis, Innsbruck, 1965

(BU2) B.BUCHBERGER Ein algorithmisches Kriterium für die Losbarkeit eines algebraischen Gleichungssystems, Aeq. Math. 4 (1970) 364-383

(BU3) B.BUCHBERGER A criterion for detecting unnecessary reductions in the construction of Gröbner bases, Proc. EUROSAM 79, L.N.Comp.Sci. 72 (1979) 3-21

(DEP) C.DE CONCINI, D.EISENBUD, C.PROCESI Hodge Algebras, Asterisque 91 (1982)

(MM1) H.M.MOELLER, F.MORA New constructive methods in classical ideal theory, J.Alg. 96 (1986)

(MM2) H.M.MOELLER, F.MORA The computation of the Hilbert function, Proc. EUROCAL 83, L.N.Comp.Sci. 162 (1983) 157-167

(MOL) H.M.MOELLER A reduction strategy for the Taylor resolution, Proc. EUROCAL 85 II, L.N.Comp.Sci. 204 (1985) 526-534

(LYU) G.LYUBEZNIK Ph.D. Thesis, Columbia Univ., 1984

(TAY) D.TAYLOR Ideals generated by monomials in an R-sequence, Ph.D.Thesis, Chicago 1960

(MIT) B.MITCHELL Theory of categories, New York, 1965

DESIGNS ARISING FROM SYMPLECTIC GEOMETRY

Lucien Bénéteau, Jacqueline Lacaze
Université Paul Sabatier
UER MIG
118, route de Narbonne
31062 TOULOUSE CEDEX - FRANCE

The Hall Triple Systems (HTSs) arose in various contexts, among others : Fischer groups, cubic hypersurface quasigroups, perfect matroids of rank $\geqslant 4$. Our results concern the HTSs of rank r whose order is 3^r (an extremal situation for the non affine HTSs of rank r, whose 3-order is always $\geqslant r$). The classification of these systems is shown to be a special case of a more general problem of symplectic geometry : the classification of the trilinear alternate forms of a finitely generated vector space. Some partial results are stated. They allow notably a complete classification of the HTSs of rank $r \leqslant 7$ whose order is minimum. For $r \leqslant 6$ one knew that there were exactly 3 such systems. We show that there are exactly five of rank 7 ones. We also determine all the HTSs of rank r and of order 3^r admitting an affine subsystem of order 3^{r-1}.

1 - INTRODUCTION : This report is divided into 4 parts. In section 2 we give a brief introduction to the Hall Triple Systems (HTSs) and to the related groups and quasigroups. There are two new statements giving precisions about the correspondence between the HTSs on one side, and the cubic hypersurface quasigroups and the Fischer groups on the other side. We refer the reader to the literature for the connections with other parts of algebra and design theory ([7,11,1]).

Further on a process of explicit construction of HTSs is recalled (section 3). This process is not canonical. But it allows to get all the non affine HTSs whose 3-order s equals the rank ρ. As usual the rank is to be understood as the minimum possible cardinal number of a generator subset. The equality $s=\rho$ corresponds to an extremal situation, the non affine HTSs obeying $s \geq \rho$, while the affine ones obey $s=\rho-1$ (see [1]). It is the classification of non affine HTSs of given rank whose order is minimal that led us to a problem of symplectic geometry.

Given some vector space V, there is a natural action of GL(V) on the set of symplectic trilinear forms of V. We shall be counting orbits in some special cases. For further investigations the most important result is some process of translation in case the field is GF(3) : there is then a one-to-one correspondence between the orbits of the non-zero forms and the isomorphism classes of some HTSs. This will be used here to obtain an exhaustive list of the HTSs of order ≤ 2187 whose ranks are $\neq 6$. We shall also classify the non affine HTSs admitting a codimension 1 affine subsystem.

2 - HALL TRIPLE SYSTEMS, MANIN QUASIGROUPS AND FISCHER GROUPS :

A Steiner Triple System is a 2-(v,3,1) design, namely it is a pair (E,L) where E is a set of "points" and L a collection of 3-subsets of E, called "lines", such that any two distinct points lie in exactly one line $\ell \in L$. The corresponding Steiner quasigroup consists of the same set E under the binary law : $E^2 \longmapsto E$; $x,y \mapsto x \circ y$ defined by $x \circ x = x$ and, whenever $x \neq y$, $x \circ y = z$, the third point of the line through x and y. The Steiner quasigroups can be algebraically characterized by the fact that the law is idempotent and symmetric. Recall that a law is said to be symmetric when any equality of the form $x \circ y = z$ is invariant under any permutation of x,y,z ; this is equivalent to the conjunction of the commutativity and the identity $x \circ (x \circ y) = y$. For a fixed set E, to endow E with a family of lines L such that (E,L) be a Steiner Triple System is equivalent to provide E with a structure of Steiner quasigroup. So in what follows we shall identify (E,L) with $(E,\circ)$.

A Hall Triple System (HTS) is a Steiner Triple system in which any subsystem that is generated by three non collinear points is an affine plane $\simeq AG(2,3)$. This additional assumption is equivalent to the fact that the corresponding Steiner quasigroup is distributive (a o (xoy)=(aox)o(aoy) identically ; see Marshall Hall Jr. [6]). Therefore the HTSs are identified with the distributive Steiner quasigroups.

Let K be a commutative field. Consider an absolutely irreducible cubic hypersurface V of the projective space $\mathbb{P}_n(K)$. Let E be the set of its non-singular K-points. Three points x,y,z of V will be said to be collinear (notation : L(x,y,z)) if there exists a line ℓ containing x,y,z such that either $\ell \subset V$ or $x+y+z=\ell.V$ (intersection cycle).

The best known case is when dim V=1, and n=2 : V is then a plane curve, it does not contain any line and overall for any x,y in E, there is exactly one point z in E such that L(x,y,z). The corresponding law $x,y \mapsto xoy=z$ is obviously symmetric. The set of the idempotent points of (E,o) is the set of flexes ; it is isomorphic to AG(t,3) with $t \leqslant 2$ (endowed with the mid-point law). Lastly for any fixed u in E, $x,y \mapsto x*y=uo(xoy)$ makes E into an abelian group. All this is classical.

Let us now consider the case dim V>1. Assume that K is infinite. We have the following fact that we mention here without all the required definitions (for a more complete account see Manin [9] pp. 46-67, especially theorems 13.1 and 13.2) :

<u>Theorem of Manin</u> : If V admits a point of "general type", then in a suitable factor set $\bar{E}$ of E, the three-place relation of collinearity gives rise to a symmetric law obeying $(aox)o(aoy)=a^2o(xoy)$ and $x^2ox^2=x^2$ identically.

As a relatively easy consequence we have :

<u>Corollary</u> : The square mapping $x \mapsto x^2=p(x)$ is an endomorphism. The set of the idempotent elements of $(\bar{E},o)$ is I=Im p ; it is a distributive Steiner quasigroup. All the fibres $A=p^{-1}(e)$ of p are isomorphic elementary abelian 2-Groups, and

$$(\bar{E}, o) \simeq I \times A \text{ (direct product).}$$

Let us say that a <u>Fischer group</u> is a group of the form $G=\langle S\rangle$ where S is a conjugacy class of involutions of G such that $O(xy)\leq 3$ for any two elements x and y from S (in other terms the dihedral group generated by any two elements of S has order ≤ 6). In case we have $O(xy)=3$ for any $x,y\in S$, $x\neq y$, G is, say, a special Fischer group. In any special Fischer group G there is just one class of involutions S_G (namely, the set of all the involutions from G), and S_G may be provided with a structure of HTS by setting $xoy=x^y=yxy(=xyx)$. We call (S_G,o) the HTS corresponding to G. This group-theoretic construction of HTSs is canonical. More precisely :

<u>Theorem</u> : Given any HTS E, the (non-empty) family $\mathfrak{F}$ of special Fischer groups whose corresponding HTS is E admits :

(i) a universal object U ; any G in $\mathfrak{F}$ is of the form $G=U/C$ where $C\subset Z(U)$.

(ii) a smallest object $I=U/Z(U)$, which is also the unique centerless element of $\mathfrak{F}$.

<u>Remarks</u> : (1) The existence of I is well-known. It can be described as the multiplication group of (E,o). (2) Very likely the foregoing properties can be generalized to general Fischer groups by considering, instead of the HTSs, the Fischer graphs related to a Fischer group [12]. We shall not discuss this question here.

3 - A PROCESS OF EXPLICIT CONSTRUCTION : Let E be a vector space over $GF(3)$ with $\dim E=n+1$. Pick up some basis $e_1,e_2,\ldots e_n,e_{n+1}$. Besides choose a non-zero sequence of $\binom{n}{3}$ elements from $GF(3)$, say :

$$\sigma=(\lambda_{ijk})_{1\leq i<j<k\leq n}$$

Define a binary law in E by setting that, if $x=\sum_{i=1}^{i=n+1} x^i e_i$ and $y=\sum_{j=1}^{j=n+1} y^j e_j$ where $x^i,y^j\in GF(3)$, then :

$$x\ o_\sigma\ y=-x-y+\Big(\sum_{1\leq i<j<k\leq n}\lambda_{ijk}(x^i-y^i)(x^jy^k-x^ky^j)\Big)\ e_{n+1}$$

Proposition : (E, o_σ) is a HTS of rank n+1. Any rank n+1 HTS of order 3^{n+1} arises in this manner.

Remark : For n=3, σ is just a non-zero scalar : ±1 ; up to isomorphism, one gets only one HTS which is the order 81 distributive Steiner quasigroup discovered by Zassenhaus (see Bol [5]).

In another connection the equalities : $t(e_i, e_j, e_k) = \lambda_{ijk}$, $1 \leqslant i < j < k \leqslant n$ determine a unique alternate trilinear form $t: V^3 \to GF(3)$ with $V = \langle e_1, e_2, \dots e_n \rangle$. The trilinear form t is linked to the law (o_σ) by :

$$t(x,y,z) = (-x) \text{ o } (yoz) - (xoy) \text{ o } (-z)$$

Conversely, starting from an alternate trilinear form t from a codimension 1 subspace $V = \langle e_1, e_2, \dots e_n \rangle$ of E, then by taking arbitrarily e_{n+1} in $E \setminus V$ one may recover $x \, o_\sigma \, y$ as follows. For any

$$z = \sum_{1 \leqslant i \leqslant n+1} z^i e_i \quad \text{in } E, \text{ let us set :}$$

$$z^{(i)} = \sum_{i < j \leqslant n} z^j e_j$$

Then : $x \, o_\sigma \, y = -x-y + \delta_1 + \delta_2 + \dots + \delta_{n-2}$ with $\delta_i = t((x^i - y^i)e_i, x^{(i)}, y^{(i)})$.

Since several different sequences $\sigma' = (\lambda'_{ijk})$ yield isomorphic HTSs, there are several such forms related to the same HTS(E, o_σ). In fact all these trilinear forms are equivalent in a sense that we are going to precise.

4 - THE PROBLEM OF CLASSIFICATION OF ALTERNATE TRILINEAR FORMS :
Consider a n-dimensional vector space V=V(n,K) over a commutative field K. Recall that a trilinear form $t: V^3 \mapsto K$, $(x,y,z) \mapsto t(x,y,z)$ is alternate (or : symplectic, skew-symmetric) if $t(\pi(x), \pi(y), \pi(z)) = \text{sign}(\pi)\, t(x,y,z)$ for every permutation π of x,y,z. Any two trilinear forms t and u are equivalent if there exists ℓ in $GL_K(V)$ such that $u(x,y,z) = t(\ell(x), \ell(y), \ell(z))$ identically. Designate as A(n,K) the

family of the non-vanishing alternate trilinear forms of V and $\tilde{A}(n,K)$ the corresponding set of equivalence classes. We shall be concerned with the classification of the trilinear forms up to equivalence in two special cases : first when there is a totally isotropic vector space, second when $n \leqslant 6$. In the first situation, n can be any integer $\geqslant 3$ but we restrict attention to the forms that vanish identically on M^3 where M is a (n-1)-dimensional subspace. When $n \leqslant 6$ we obtain an almost complete classification, and in fact a complete one in case K=GF(3). Now any result concerning the special case K=GF(3) can be interpreted in terms of HTSs in view of the following result :

<u>Theorem 4.1</u> (translation theorem) : There is a one-to-one correspondence between $\tilde{A}(n, GF(3))$ and the isomorphism classes of HTSs of rank n+1 whose order is 3^{n+1}.

This translates the problem of the classification of non affine HTSs of maximal rank into an exterior algebraic classification problem. We shall study two special cases where the number of classes can be specified. As a first example let us give a couple of statements consisting first of a classification theorem of symplectic geometry and second of an application to the case K=GF(3) through the translation theorem.

<u>Theorem 4.2</u> : For any commutative field K the elements of A(n,K), $n \geqslant 3$, admitting a totally isotropic codimension 1 subspace form $[^{n-1}/2]$ complete equivalence classes. If n is odd there is only one such class for which V is non singular ; if n is even there is none.

<u>Corollary</u> : The rank n+1 HTSs of order 3^{n+1}, $n \geqslant 3$, admitting an affine sybsystem of order 3^n form exactly $[^{n-1}/2]$ complete isomorphism classes. If n is odd there is exactly one such system without non trivial affine direct factor ; if n is even there is none.

For instance if n=9, there are 4 rank 10 HTSs of order 3^{10} admitting an affine order 3^9 subspace ; they can be described in the process of section 3 by setting : $\lambda_{123}=1$, $\lambda_{145}=\alpha$, $\lambda_{167}=\beta$, $\lambda_{189}=\gamma$ and, for $\{i,j,k\} \notin \{\{1,2,3\}, \{1,4,5\}, \{1,6,7\}, \{1,8,9\}\}$ $\lambda_{ijk} = 0$, where α,β,γ are 0 or 1. The non singular case arises when one takes $\alpha=\beta=\gamma=1$.

Theorem 4.3 : For any commutative field K, $|\tilde{A}(3,K)|=1=|\tilde{A}(4,K)|$ and $|\tilde{A}(5,K)|=2$. Besides $|\tilde{A}(6,GF(3))|=5$.

Corollary : Up to isomorphism there are exactly (i) one rank 4 (resp. 5) HTS of order 3^4 (resp. 3^5), (ii) two rank 6 HTSs of order 3^6, and (iii) five rank 7 HTSs of order 3^7.

Remark : The link between theorem 4.3 and the corollary follows from the translation theorem. It gives a quick proof of (i) and (ii) that had been previously established [6,8,1,2,3,4,10].

As an intermediate stage in the proof of theorem 4.3 one gets the following result which is of some interest because it is valid for an arbitrary commutative field K :

Lemma : Any element t from A(6,K) satisfies two or none of the following conditions : (i) there exists at least a non singular 3-dimensional subspace P such that $(P^{\perp})^{\perp} \cap P \neq \{0\}$; (ii) there exists at least a singular codimension 1 subspace. The subsets of A(6,K) consisting of the trilinear forms satisfying these conditions is the set-union of exactly four classes.

It turns out that, if K=GF(3), then all the skew-symmetric trilinear forms which do not satisfy the condition (i) form a unique class ; the HTS related to this class can be described in the process of section 3 by setting : $\lambda_{123}=-1$, $\lambda_{145}=1=\lambda_{246}=\lambda_{356}$, and, for $\{i,j,k\} \notin \{\{1,2,3\}, \{1,4,5\}, \{2,4,6\},\{3,5,6\}\}$, $\lambda_{ijk}=0$.

BIBLIOGRAPHY.

[1] L. BENETEAU : Topics about Moufang loops and Hall triple systems, Simon Stevin 54(1980) 107-124.

[2] L. BENETEAU : Une classe particulière de matroïdes parfaits, in "Combinatorics 79", Annals of Discrete Math. 8 (1980) 229-232.

[3] L. BENETEAU : Les Systèmes Triples de Hall de dimension 4, European J. Combinatorics (1981) 2, 205-212.

[4] L. BENETEAU : Hall Triple systems and related topics. Proc. Interna. Conf. on Combinatorial Geometries and their applications, 1981 ; Annals of Discrete Math. 18 (1983) 55-60.

[5] G. BOL : Gewebe und Gruppen, Math. Ann. 114 (1937) 414-431 ; Zbl. 16, 226.

[6] Marshall HALL Jr. : Automorphisms of Steiner Triple systems, IBM J. Res. Develop. (1960), 406-472. MR 23A#1282 ; Zbl.100, p.18.

[7] Marshall HALL Jr. : Group theory and block designs, in Proc. Intern. Conf. on the Theory of Groups, Camberra 1965, Gordon and Breach, New-York, MR 36#2514 ; Zbl. 323.2011.

[8] T. KEPKA : Distributive Steiner quasigroups of order 3^5, Comment. Math. Univ. Carolinae 19,2 (1978) ; MR 58 # 6032.

[9] Yu. I. MANIN : Cubics forms, North-Holland Publishing Company, Amsterdam-London ; American Elsevier Publishing Company, Inc. New-York (1974).

[10] R. ROTH, RAY-CHAUDHURI D.K. : Hall Triple Systems and Commutative Moufang Exponent 3 Loops : the case of Nilpotence Class 2, J. Comb. Theory A 36 (1984) 129-162.

[11] H.P. YOUNG : Affine triple systems and matroïd designs, Math. Z. 132 (1973) 343-366. MR. 50 # 142.

[12] F. ZARA : Une caractérisation des graphes associés aux groupes de Fischer (submitted for publication in European Journal of Combinatorics).

DISTANCE - TRANSITIVE GRAPHS AND THE PROBLEM OF MAXIMAL SUBGROUPS OF SYMMETRIC GROUPS

A. ASTIE-VIDAL and J. CHIFFLET

Laboratoire MLAD
Université Paul Sabatier
TOULOUSE

SUMMARY :

We give a necessary and sufficient condition for the automorphism group of a distance-transitive graph to be a maximal unitransitive subgroup of the symmetric group (theorem 1). Then we use the necessary condition of theorem 1 to determine the automorphism groups of a class of graphs (theorem 2). After that, we use the sufficient condition of theorem 1 to determine a class of maximal unitransitive subgroups of the symmetric group S_{m^d} .

I - DISTANCE - TRANSITIVE GRAPHS

All graphs considered will be finite, undirected, connected and with no loops or multiple edge, we denote the distance between two vertices x and y by $d(x,y)$.

Let $G = (X,U)$ be a graph and let AutG be the automorphism group of G.

DEFINITIONS

- G is vertex-transitive if $\forall x,y \in X$, $\exists \sigma \in \text{Aut } G : \sigma(x)=y$.
- G is edge-transitive if $\forall [x,y] \in U$, $\forall [x',y'] \in U$, $\exists \sigma \in \text{Aut } G$: $\sigma(x)=x'$, $\sigma(y)=y'$.
- G is distance-transitive if for all vertics x,y,x',y' such that $d(x,y)=d(x',y')$, there is a $\sigma \in \text{Aut } G$ such that : $\sigma(x)=x'$, $\sigma(y)=y'$.
- A group is unitransitive if it is transitive but not 2-transitive. Clearly, if G is a vertex-transitive graph, we have (i) or (ii)
  - (i) Aut G is unitransitive,
  - (ii) G is the complete graph or the empty graph.

Let $G=(X,U)$ be a distance-transitive graph of diameter d, and for every $k=0,1,\ldots\ldots,d$; let

$$U_k = \{(x,y) \in X \times X : d(x,y) = k\} .$$

From the definition of a distance-transitive graph, it is clear that the orbits of the group of automorphisms of G on XxX (i.e the 2-orbits) are the sets :

$$U_0, U_1, \ldots\ldots, U_d$$

Let $G_k=(X,U_k)$; $k=1,\ldots\ldots,d$, be the graph with vertex-set X and edge-set U_k (note that $G_1=G$).

Clearly the G_k's are vertex and edge-transitive graphs.

By extension, for any non-trivial subset J of $\{1,2,\ldots\ldots,d\}$, (by non trivial we mean that J is not the empty set nor the whole set $\{1,2,\ldots\ldots,d\}$), we define the graphs $G_J=(X,U_J)$ with

$$U_J=\{(x,y):d(x,y)\in J\} \quad (\text{i.e } U_J=\bigcup_{k\in J} U_k)$$

REMARK :

The number of such graphs G_J is 2^d-2 and when $J=\{k\}$ is a 1-subset we have $G_{\{k\}}=G_k$, $\forall\ k=1,\ldots\ldots,d$.

The graphs G_J are vertex-transitive (not necessarily edge-transitive) graphs such that $\text{Aut } G \leq \text{Aut } G_J$, $\forall$ J non trivial subset of $\{1,\ldots\ldots,d\}$.

THEOREM 1 :

Let G be a distance-transitive graph of order n and diameter d.

Aut G is maximal as a unitransitive subgroup of S_n if and only if :

$$\text{Aut } G = \text{Aut } G_J \ ; \ \forall J,\ \emptyset \neq J \subsetneq \{1,\ldots\ldots,d\}$$

Proof :

- The maximality of Aut G as a unitransitive subgroup of S_n, together with the fact that $\text{Aut } G \leq \text{Aut } G_J$, implies that $\text{Aut } G = \text{Aut } G_J$, since from $J \neq \{1,\ldots,d\}$ we have that $\text{Aut } G_J$ is not 2-transitive.

- Conversely, suppose that there exists a unitransitive subgroup H of S_n, such that : $\text{Aut } G < H < S_n$. Then, the 2-orbits of H are unions of 2-orbits of Aut G. That is $\exists J \subset \{1,\ldots\ldots,d\}$ such that U_J is a 2-orbit of H ; $J \neq \emptyset$ and $J \neq \{1,2,\ldots,d\}$ since H is not 2-transitive. Then, $\text{Aut } G < H \leq \text{Aut } G_J < S_n$, contradicts the fact that

$$\text{Aut } G = \text{Aut } G_J,\ \forall J .$$

This theorem was proved, with a little different formulation in (1) (Th. I-2-2. p 59). Now, we give two applications of this theorem, each uses one side of the equivalence : The first one uses known results on the maximality of a class of subgroups of symmetric group to determine the automorphism group of a family of graphs, the results obtained were not yet obtained in another way. The second application consists in determining automorphisms of graphs and deducing maximal subgroups of symmetric groups.

II - THE DISTANCE-TRANSITIVE GRAPH $G\binom{m}{d}$

Define $G\binom{m}{d} = (X,U)$ to be the following graph :

- X it the set of all the d-subsets of a m-set E, thus X has cardinality

$$\binom{m}{d} = \frac{m!}{d!(m-d)!}$$

- Two elements x and y of X are joined by an edge of G if and only if the cardinality of their intersection (as subsets of E) is d-1 ; (that is $d-|x \cap y|=1$).

Clearly, $G\binom{m}{d}$ is a distance-transitive graph of order $\binom{m}{d}$ and diameter d.

From Enomoto (7), it is known that :

$$\text{Aut } G \cong S_m, \text{ if and only if } m \neq 2d$$

This result was proved in the context of matroid theory in (1). In (9), (10) and (11), it is proved that S_m is maximal as unitransitive subgroup of $S_{\binom{m}{d}}$, if $m \neq 2d-1, 2d, 2d+1$, or if $m = 5$, $d = 2$.

From the preceding theorem, the following result is straightforward.

Theorem 2 :

For any non trivial subset J of $\{1,2,\ldots,d\}$, let $G_J\binom{m}{d} = (X,U_J)$ be the following graph :X is the set of all the d-subsets of an m-set E.

x and y, belonging to X, satisfy :

$(x,y) \in U_J$ if and only if $d - |x \cap y|$ belongs to J, Then we have :

$$\text{Aut } G_J\binom{m}{d} \cong S_m \text{ if } m \neq 2d-1,\ 2d,\ 2d+1 \text{ or if } m=5,\ d=2$$

The particular case of our theorem, m=5, d=2, proves that the automorphism group of the Petersen graph is isomorphic to S_5 (m=5, d=2, $G_2\binom{5}{2}$ is the Petersen graph).

III - THE DISTANCE-TRANSITIVE GRAPH $G[m^d]$

LET $G[m^d] = (X,U)$ be the graph :

- X is the set of all the d-uple from a m-set E ($X=E^d$),
- two elements x and y of X are joined by an edge of $G[m^d]$ if and only if x differs from y by exactly one coordinate.

Clearly, $G[m^d]$ is a distance-transitive graph of order m^d and diameter d.

It is known that Aut $G[m^d] \cong S_d \sim S_m$ (wreath-product of the symmetric group of degree d with the one of degree m).This result was first proved in (8), also proved in the context of matroid theory in (1), and of association schemes in (5). In fact, the sets $U_0,U_1,\ldots,U_d$ associated with this particular distance-transitive graph are the classes of a Hamming scheme on X, and, we use now the parameters of this association scheme :

$$p_{ij}^k = \text{card } \{x \in X : (x,y) \in U_i,\ (x,z) \in U_j\}, (y,z) \in U_k \ ;\ 0 \leq i,j,k \leq d$$

In the following, we say that the p_{ij}^k are the parameters of $G[m^d]$.

LEMMA 1 :

Let $p_{ij}^k (0 \leq i,j,k \leq d)$ denote the parameters of $G[m^d]$ and $\overline{p}_{ij}^k$ $(0 \leq i,j,k \leq d-1)$ the parameters of $G[m^{d-1}]$, we have :

$$p_{ij}^k = \overline{p}_{ij}^k + (m-1)\,\overline{p}_{i-1,j-1}^k \quad : 0 \leq k \leq d-1 \ ; \ 1 \leq i,j \leq d-1$$

Let $\overline{x} = (x_1, x_2, \ldots, x_{d-1})$ and $\overline{y} = (y_1, y_2, \ldots, y_{d-1})$ be vertices of $G[m^{d-1}]$ such that $d_{G[m^{d-1}]}(\overline{x},\overline{y}) = k \quad (0 \leq k \leq d-1)$.

Denote $x = (x_1, \ldots, x_{d-1}, x_d)$ and $y = (y_1, \ldots, y_{d-1}, x_d)$ two vertices of $G[m^d]$.

Clearly, $d_{G[m^d]}(x,y) = d_{G[m^{d-1}]}(\overline{x},\overline{y}) = k$.

Denote $f_{ij}^k = \{z \in E^d : d(x,z) = i,\ d(y,z) = j\}$; thus $p_{ij}^k = \text{card } f_{ij}^k$.

And $\overline{f}_{ij}^k = \{\overline{z} \in E^{d-1} : d(\overline{x},\overline{z}) = i,\ d(\overline{y},\overline{z}) = j\}$; thus $\overline{p}_{ij}^k = \text{card } \overline{f}_{ij}^k$.

For any $z \in f_{ij}^k$, we have different possibilities :

if $\quad z = (z_1, \ldots, z_{d-1}, z_d)$ and $\overline{z} = (z_1, \ldots, z_{d-1})$

(i) $\quad \overline{z} \in \overline{f}_{ij}^k$ and $z_d = x_d$

(ii) $\quad \overline{z} \in \overline{f}_{i-1,j-1}^k$ and $z_d \neq x_d$, that is $(m-1)$ choices for z_d, then the result is straightforward : $p_{ij}^k = \overline{p}_{ij}^k + (m-1)\overline{p}_{i-1,j-1}^k$.

Proposition 1 :

When $m \geq 5$, the parameters of $G[m^d]$ satisfy :

$$p_{ii}^0 > p_{ii}^1 > \ldots > p_{ii}^{k_1} > 0 \ ; \ k_1 = \inf\{2i,d\}$$

$$p_{ii}^k = 0, \quad \forall\, k \geq k_1 + 1 .$$

clearly $p_{ii}^k > 0,\ \forall\, k \leq k_1$ and $p_{ii}^k = 0,\ \forall\, k \geq k_1 + 1$.

The proof is by induction on d.

Let d=2 : the parameters of $G[m^d]$ are :

$$p_{00}^0 = 1$$

$$p_{11}^0 = 2(m-1) \qquad p_{11}^1 = (m-2) \qquad p_{11}^2 = 2$$

$$p_{22}^0 = (m-1)^2 \qquad p_{22}^1 = (m-1)(m-2) \quad p_{22}^2 = (m-2)^2$$

and they satisfy the property, when $m \geq 5$.

Let us suppose that the property is true for the parameters $\overline{p}_{ii}^k$ of $G[m^{d-1}]$ $(0 \leq i,\ k \leq d-1)$ that is :

$$\overline{p}_{ii}^0 > \overline{p}_{ii}^1 > \ldots > \overline{p}_{ii}^{\overline{k}_1} > 0 \quad \overline{k}_1 = \inf\ \{2i,d-1\}$$

$$\overline{p}_{ii}^k = 0\ , \quad \forall\, k \geq \overline{k}_1 + 1$$

From the preceding lemma, we know that :

$$p_{ij}^{k} = \bar{p}_{ij}^{k} + (m-1)\,\bar{p}_{i-1,j-1}^{k} \;;\quad 0 \leqq i,j,k \leqq d-1,\ i,j \neq 0$$

From this, and the induction hypothesis, we deduce :

$$p_{ii}^{0} > p_{ii}^{1} > \ldots > p_{ii}^{\bar{k}_1} > 0 \quad (0 \leqq i \leqq d-1,\ \bar{k}_1 = \inf\{2i, d-1\})$$

(i) if $i \leqq d-1/2$, we have $2i \leqq d-1 < d$, thus $\bar{k}_1 = k_1 = 2i$, and the result follows.

(ii) if $d/2 \leqq i \leqq d-1$, we have $d-1 < d \leqq 2i$, thus $\bar{k}_1 = d-1$, $k_1 = d$; and we have to prove $p_{ii}^{d-1} > p_{ii}^{d} > 0$.

Simple counting arguments give :

$$p_{ii}^{d} = \binom{d}{i}\binom{d-i}{i}(m-2)^{d-2i}$$

$$p_{ii}^{d-1} = \binom{d-1}{i}\binom{d-i-1}{i}(m-1)(m-2)^{d-2i+1} + \binom{d-1}{i-1}\binom{d-i}{i-1}(m-2)(m-2i-1)$$

thus $p_{ii}^{d-1} > p_{ii}^{d}$ reduces to

(1) $\qquad i^2 m^2 - 2i(d+1)m + 2i + d(d+1) > 0$.

The discriminant of this trinom in m is :

$\Delta = i^2(d+1-2i)$, and we are in the case $d \leqq 2i$.

- $d < 2i-1 \iff \Delta < 0$ and (1) is true .
- $d = 2i-1 \iff \Delta = 0$, the unique root is $m = d+1/i = 2$, thus for $m \geqq 5$, (1) is true.
- $d = 2i \iff \Delta > 0$, the roots are : $m' = 2$, $m'' = 2(i+1)/i \leqq 4$ and for $m \geqq 5$, (1) is true.

To achieve the proof of the theorem, we have to treat a last case : $i = d$

(iii) $\quad i = d$; we have to prove :

$$p_{dd}^{0} > p_{dd}^{1} > \ldots > p_{dd}^{d} > 0$$

for $0 \leqq k \leqq d$, we have $p_{dd}^{k} = (m-1)^{d-k}(m-2)^{k}$ thus $p_{dd}^{k} > p_{dd}^{k+1} \iff m-1 > m-2$ which is always true.

This last result (iii) is also a consequence of a more general result on matroids, proved in (1).

<u>Proposition 2</u>

Let $G_i[m^d]$ be the graph with vertex set E^d, ($|E| = m$) ; two vertices being joined if and only if they differ by exactly i coordinates ($1 \leqq i \leqq d$). Then, if $m \geqq 5$, we have :

$$\mathrm{Aut}\, G_i[m^d] \cong S_d \sim S_m$$

At the beginning of § III, we have seen that Aut $G[m^d] \cong S_d \sim S_m$.

The graph $G[m^d]$ is, in fact, $G_1[m^d]$, thus Aut $G_1[m^d] \cong S_d \sim S_m$.

An automorphism of $G_1[m^d]$ preserves the distances in this graph, and so, is an automorphism of $G_i[m^d]$. Thus, we have :

$$S_d \sim S_m \cong \text{Aut } G_1[m^d] \leq \text{Aut } G_i[m^d] \quad ; \forall\, i : 1 \leq i \leq d .$$

And we have to prove that $\text{Aut}G_i[m^d] \leqq \text{Aut } G_1[m^d]$ $(m \geqq 5)$, to achieve the proof of the theorem :

- Let $\sigma \in \text{Aut } G_i[m^d]$. We denote $f_{ii}(x,y) = \{z \in E^d : d(x,z)=i,\ d(y,z)=i\}$

Clearly $\sigma(f_{ii}(x,y)) = f_{ii}(\sigma(x),\sigma(y))$, thus card $f_{ii}(x,y) =$ card $f_{ii}(\sigma(x),\sigma(y))$

Thas is $p_{ii}^{d(x,y)} = p_{ii}^{d(\sigma(x),\sigma(y))}$. But from the proposition 1, we have that, if $d(x,y) = k \leqq k_1 = \inf(2i,d)$, this implies $d(x,y) = d(\sigma(x),\sigma(y)) = k$.

And we have proved that :

$$\forall\, k : 1 \leqq k \leqq k_1 = \inf\{2i,d\} \ , \quad \text{Aut } G_i[m^d] \leqq \text{Aut } G_k[m^d]$$

and in particular $\quad \text{Aut } G_i[m^d] \leqq \text{Aut } G_1[m^d]$. C.Q.F.D.

Recall that our aim is to apply theorem 1 to the distance-transitive graphs $G[m^d]$; so as to prove the maximality of $S_d \sim S_m$ as a unitransitive subgroup, in the symmetric group S_{m^d}.

We have proved that, for $m \geqq 5$, we have Aut $G_i[m^d] \cong S_d \sim S_m (1 \leqq i \leqq d)$ but we have to prove that Aut $G_J[m^d] \cong S_d \sim S_m$ for any non-trivial subset J of $\{1,2,\ldots,d\}$ and not only for 1-subsets $J=\{i\}$. We are able to prove this result only for d=2,3,4,5.

First, we give a preliminary lemma :

LEMMA 2 :

Let $m \geqq 5$, and H be a unitransitive subgroup of S_{m^d} containing strictly $S_d \sim S_m$. Let us denote $J_1, J_2, \ldots, J_s$ the partition of $\{1,2,\ldots,d\}$ such that $U_{J_1}, U_{J_2}, \ldots, U_{J_s}$ are the 2-orbits of H; this partition has to satisfy :

(i) $\quad 2 \leqq \text{card } J_\ell \leqq d-2 \ ; \ \forall\, \ell = 1,\ldots,s$

(ii) $\quad \forall\, (\ell,m,r) \in \{1,2,\ldots,s\}^3 : \displaystyle\sum_{(i,j) J_\ell \times J_m} p_{ij}^k = \tilde{p}^r_{\ell m} \quad \forall\, k \in J_r$

(iii) $\quad \forall\, k > k' \ ; \ k,k' \in J_r$, we have $\displaystyle\sum_{\substack{i<j \\ i,j \in J_\ell}} p_{ij}^k = \sum_{\substack{i<j \\ i,j \in J_\ell}} p_{ij}^{k'}$

if $k' > \sup\{2i, i \in J_\ell\}$ and $\displaystyle\sum_{\substack{i<j \\ i,j \in J_\ell}} p_{ij}^k > \sum_{\substack{i<j \\ i,j \in J_\ell}} p_{ij}^{k'}$ in other cases .

Proof :

(i) follows from the fact that :

$$H \leq \bigcap_{i=1}^{s} \text{Aut } G_{J_i}$$; and if card $J_i = 1$, we have from proposition 2

that : Aut $G_{J_i} = S_d \sim S_m$, which implies that $H = S_d \sim S_m$, contradicts the hypothesis of the lemma.

(ii) follows from the fact that the 2-orbits of H satisfy the same regularity conditions that the 2-orbits of Aut G.

(iii) In cas ℓ=m, (ii) can be written :

$$2 \sum_{\substack{i<j \\ i,j \in J_\ell}} p_{ij}^{k} + \sum_{i \in J_\ell} p_{ii}^{k} = 2 \sum_{\substack{i<j \\ i,j \in J_\ell}} p_{ij}^{k'} + \sum_{i \in J_\ell} p_{ii}^{k'}$$

$k>k'$ ======> $\sum_{i \in J_\ell} p_{ii}^{k} < \sum_{i \in J_\ell} p_{ii}^{k'}$ except in the case where these

parameters are all null, that is if $k' > \sup \{2i, i \in J_\ell\}$ and the result follows.

Proposition 3 :

If $m \geq 5$, and d=2,3,4,5, we have : Aut $G_J[m^d] \cong S_d \sim S_m$, for any non trivial subset J of $\{1, \ldots, d\}$.

(1) d=2 : Any non trivial subset of {1,2} is a 1-subset and the result follows from the proposition 2 and the lemma 2.

(2) d=3 : Any partition of {1,2,3} into non trivial subsets contains at least one 1-subset and the result follows in the same way as in (1).

(3) d=4 : Any partition of {1,2,3,4} into non trivial subsets of cardinality ≥ 2, has exactly two classes of cardinality 2 :

$$J_1 = \{i,j\}, \quad J_2 = \{k,\ell\} \quad (\{i,j,k,\ell\} = \{1,2,3,4\})$$

(i) $J_1 = \{1,2\}$ $J_2 = \{3,4\}$

$p_{12}^{1} = 3(m-1)$, $p_{12}^{2} = 2(m-2)$ thus $p_{12}^{1} > p_{12}^{2}$ which contradicts lemma 2 (with k=2, k'=1, $J_\ell = J_r = \{1,2,3\}$).

(ii) $J_1 = \{1,3\}$, $J_2 = \{2,4\}$ $p_{13}^{1} = 0$ $p_{13}^{3} = 0$ contradicts lemma 2

(with $J_\ell = J_r = \{1,3\}$, $k' = 1$, $k = 3$).

(iii) $J_1 = \{2,3\}$ $J_2 = \{1,4\}$ $p_{23}^2 = 4(m-1)(m-2)$

$p_{23}^3 = 3(m-2)^2 \Longrightarrow p_{23}^2 > p_{23}^3$ contradicts lemma 2

(with $J_\ell = J_r = \{2,3\}$, $k'=2$, $k=3$).

(4) $d=5$: Any partition of $\{1,2,3,4,5\}$ into non-trivial subsets of cardinality ≥ 2 has exactly two classes of cardinality 2 and 3 respectively.

Thus we have $\binom{5}{2} = 10$ possibilities. It would be too long to write here the detailed proof. Each of the ten cases can be solved in a similar way as for $d=4$, using lemma 2.

THEOREM 3 :

If $m \geq 5$ and $d=2,3,4,5$; the wreath-product of symmetric groups : $S_d \sim S_m$ is maximal as a unitransitive subgroup of $S_m d$.

This is an immediate consequence of theorem 1 and proposition 3.

REFERENCES

(1) - A. ASTIE-VIDAL Sur certains groupes d'automorphismes de graphes et d'hypergraphes.
Thèse Université de Toulouse (Juin 1976).

(2) - A. ASTIE-VIDAL Groupes d'automorphismes des graphes d'intersection aux bases d'un matroïde.
Coll. Math. discretes : codes et hypergraphes.
Bruxelles 1978. Cahiers C.E.R.O. 20.3-4 - 1978

(3) - A. ASTIE-VIDAL Factor groups of the automorphism group of a matroid basis graph with respect to the automorphism group of the matroid.
Discrete maths 32 (1980) - 217-224.

(4) - A. ASTIE-VIDAL The automorphism group of a matroid.
Annals of discrete maths 9 (1980) - 205-216.

(5) - P. BONNEAUD Thèse - Paris (1984).

(6) - J. CHIFFLET Schémas d'association et groupes d'automorphismes de graphes et hypergraphes.
Thèse de 3e cycle - Toulouse (mars 1984)

(7) - H. ENOMOTO Characterization of families of finite permutation groups by the subdegrees I.
J. Fac. Sciences University of Tōkiō, Ia. Vol. 19 N° 1, (129-135) - (1972).

(8) - H. ENOMOTO Characterization of families of finite permutation groups by the subdegrees II.
J. Fac. Sciences, University of Tokio, Ia, VoL. 19 N° 1 (1-11) - (1973).

(9) E. HALBERSTADT On certain maximal subgroups of symmetric or alternating groups, Math. Z, 151 - 117-125 (1976) .

(10) L.A. KALOUJNINE, N.H. KLIN On certain maximal subgroups of symmetric or alternating groups. Math. U.S.S.R. Sbornik, vol. 16, N 1 (1972).

(11) J. SKALBA On the maximality of S_n in $S_{\binom{n}{k}}$.
J. of algebra 75, 158-174. (1982).

CAN A FAST SIGNATURE SCHEME WITHOUT SECRET KEY BE SECURE

Paul CAMION
INRIA
Domaine de Voluceau
Rocquencourt
B.P. 105
78153 Le Chesnay Cedex
FRANCE

ABSTRACT

Another title could have been "A probabilistic factorization algorithm in GL(2,p)". However, the problem is to calculate a fast and short signature associated with a plaintext inscribed on an erasable support. The signature should be written down in a book accompanying the record in order that it could be checked anytime that the latter has not been changed. J. BOSSET [1] suggest such a scheme together with an algorithm for computing a signature. The 64 characters needed for the plaintext are identified with a subset of GL(2,p), $p = 997$. The signature is the product of the matrices corresponding to the plaintext characters taken in the order where they appear. Such a scheme could be broken if it is possible to factorize an element of GL(2,p) into $t = 16\ r$ factors, each one in a subset $\mathbf{U}_i$ of GL(2,p) of size 64 , $i = 1,\dots,t$. We here assume one hypothesis only on uniform probability distributions of random variables defined on product sets $\mathbf{V}_j = \mathbf{U}_{jr+1} \times \dots \times \mathbf{U}_{(j+1)r}$, $j = 0,\dots,15$. In consideration on which, a probabilistic factorization algorithm in GL(2,p) is introduced.

It is shown that for $p = 10{,}007$, drawing according to a uniform probability distribution a sequence of 11,952 elements in each $\mathbf{V}_j$ provides the whole needed material to factorizing with a probability of success of at least 97%. The most expensive operation in the algorithm is sorting each of the sequences.

UN SCHEMA DE SIGNATURE COURTE ET RAPIDE N'UTILISANT PAS DE CLE SECRETE PEUT-IL ETRE FIABLE ?

Résumé

Nous étudions une solution proposée au problème d'une signature courte, calculable rapidement, associée à un texte inscrit sur un support effaçable. La (ou les) signature serait inscrite dans un livre accompagnant l'enregistrement de façon à pouvoir vérifier à tout moment que ce dernier n'a pas été modifié. J. BOSSET [1] propose un tel schéma et un algorithme de signature. Il identifie l'ensemble des 64 caractères utilisés pour le texte à un sous-ensemble de $GL(2,p)$, $p = 997$. La signature est le produit des matrices correspondant aux caractères du texte effectué dans l'ordre où ceux-ci apparaissent. On peut casser ce schéma s'il est possible de factoriser un élément de $GL(2,p)$ en $t = 16\,r$ facteurs, chacun dans un sous-ensemble $\mathbf{U}_i$ de taille 64 de $GL(2,p)$, $i = 1,\ldots,t$. Nous faisons ici une seule hypothèse de distributions de probabilités uniformes de variables aléatoires définies sur les ensembles produits $\mathbf{V}_j = \mathbf{U}_{jr+1} \times \ldots \times \mathbf{U}_{(j+1)r}$, $j = 0,\ldots,15$. Moyennant ceci, un algorithme probabiliste de factorisation dans $GL(2,p)$ est introduit.

On montre que pour $p = 10.007$, un tirage selon une loi uniforme d'une suite de 11.952 éléments dans chacun des $\mathbf{V}_j$ fournit tout le matériel nécessaire à la factorisation avec une probabilité de succès supérieure à 0,97. L'opération la plus coûteuse est un tri de chacune de ces suites.

INTRODUCTION

Le problème d'une signature courte, calculable rapidement, associée à un texte inscrit sur support effaçable nous a été exposé par M. GORIN, responsable d'un groupe d'étude de la commission de l'ordinateur de compensation. Pratiquement, il s'agit de s'assurer après le transfert, par exemple d'une bande magnétique, que les données inscrites n'ont pas été modifiées entre la source et l'arrivée. La solution proposée par M. BOSSET [1] avait attiré l'attention de M. GORIN pour son intérêt pratique remarquable. En résumé, il s'agit de sectionner le texte, c'est-à-dire la suite des caractères inscrits, en blocs de, disons, 10.000 caractères et d'associer à chacun de ces blocs un nombre d'au moins 20 chiffres décimaux (*) nommé signature

---

(*) En vérité, M. BOSSET suggérait une signature de 12 chiffres. Compte-tenu de la possibilité d'attaques probabilistes valables pour tout schéma de signature, y compris pour un schéma utilisant des quantités secrètes, il semble raisonnable aujourd'hui de ne pas se fier à une signature de moins de 20 chiffres décimaux.

et calculé sur la donnée du bloc de telle sorte que tout contrôle ultérieur ferait apparaître une signature totalement différente de l'originale si une modification, même mineure, des caractères du bloc était survenue. On veut évidemment être assuré de la sécurité du test que constitue l'identité de la signature calculée à la signature originale. Le test prend d'ailleurs le nom de certification. Mais il faut aussi que ce calcul soit très rapide. Pour fixer les idées, il faut pouvoir calculer les signatures relatives à 175.000 caractères en une seconde d'unité centrale de gros ordinateur dans le cas de compensations bancaires.

M. BOSSET propose également d'être en mesure de calculer la signature d'une suite de signatures concaténées. De cette façon on pourrait "résumer" plusieurs millions de caractères en quelques centaines de chiffres décimaux qui occuperaient peu d'espace sur un support non effaçable où l'on pourrait vérifier, comme c'est l'usage pour les livres comptables, qu'il n'y a eu ni ratures ni rajouts. Si une certification devait faire défaut on pourrait demander à la source une retransmission.

D'autres applications d'un tel schéma de signature courte et rapide sont proposées par M. BOSSET. Il s'agit essentiellement, dit-il, de prouver qu'une information n'a pas été altérée depuis sa création. Le schéma est donc applicable à tout fichier informatique dont les signatures seraient inscrites au "grand livre", et exploitées à toute édition pour certification.

M. BOSSET [1] propose dans son article un schéma particulier de signature et l'objet du présent article est de démontrer que ce schéma n'est pas fiable. Le problème reste donc posé de trouver un schéma de signature courte, rapide et qui n'imposerait pas la gestion de clés secrètes.

I. - THE GENERAL CONSIDERED SCHEME OF SIGNATURE

1.1. - The mapping

Let X be an alphabet, say, of 64 characters. Let k be an integer which will be the **fixed length** of a plaintext . Then k may be choosen once for all, say, from 1,000 to 10,000. A plaintext is considered to be any element from X^k. We then define a mapping σ from X^k onto the set $\{0,1,...,9\}^\ell$ where ℓ is an integer. For instance $\ell = 24$. Thus for b in X^k, then $\sigma(b) = s$ is the signature of b. The signature is then an integer of 24 decimal digits.

1.2. - The aim of the considered signature scheme

The plaintext is supposed to be easily erasable or possibly changed to another one. We then intend to compute the sequence of signatures corresponding to the given sequence of plaintexts. The sequence of signature is written down in a book and we assume that a fake is as unlikely to be forged as for usual writings. Hence if it was practically impossible to change a plaintext to another one with the same signature, then the only needed precaution to be taken for guaranteeing data recorded, say, on a magnetic tape, would be to join with it a book containing references to all plaintexts and the corresponding signatures.

1.3. - The requirements

1.3.1. - The signature should be easily computed

1.3.2. - The probability that two plaintexts have the same signature should be close to $10^{-\ell}$.

1.3.3. - It should be practically impossible to change the plaintext to another one having the same signature, by any means.

1.4. - A suggested signature

1.4.1. - Computing in the group GL(2,p)

J. BOSSET [1] suggest the above general scheme. He also suggest a particular function σ and the aim of the present paper is to show that σ does not fulfil requirement 1.3.3.. The alphabet X is identified with a subset of size 64 of GL(2,p) for $p = 997$. Hence a plaintext b is a sequence $(b_1,\ldots,b_k)$ of matrices from X and $\sigma(b)$ is nothing else but the product $b_1\ldots b_k$.

1.4.2. - Examining the requirements

Obviously 1.3.1. is fulfilled. The author carefully chooses the subset X of GL(2,p) in view of 1.3.2. In regard to 1.3.3., the author observes that forging a false needs being able to factorize and s from GL(2,p) into matrices from the samll subset X. We will set the problem of forging a false by such a factorization in the next paragraph. Then in section 2 we solve the factorization problem by means of a probabilistic algorithm.

1.4.3. - Forging a fake by factorization

We denote by G the group $GL(2,p)$ of invertible 2 by 2 matrices with entries in the finite field $\mathbb{F}_p$, p a prime. We will keep $|X| = 64$ as suggested by J. BOSSET, but we make $p = 10{,}007$ to make the problem somewhat harder. An integer t will appear in the following. For the algorithm, t will be a multiple of 16. We suggest $t = 48$ but $t = 64$ could possibly provide better results for the required statistical tests.

1.4.3.1. - The considered fraud

Given a plaintext

$$(b_1,b_2,\ldots,b_{i_1},\ldots,b_{i_t},\ldots,b_k) \in X^k$$

of which the signature is s :

$$b_1b_2 \ldots b_k = s ,$$

and where $i_1,i_2,\ldots,i_t$ is any given subset of size t of $[0,k]$, the fraud is as follows. I change the whole plaintext for a new one at the exception of characters in positions $i_1,\ldots,i_t$. The new plaintext is of my own choice. Then it will be possible to adapt the values of matrices $b_{i_1},\ldots,b_{i_t}$ in order that the new plaintext writes

$$(b_1',\ldots,b'_{i_1-1},x_1,b'_{i_1+1},\ldots,b'_{i_2-1},x_2,b'_{i_2+1},\ldots,b'_{i_t-1},x_t,b'_{i_t+1},\ldots;b_k')$$

having the same signature s.

1.4.3.2. - The factorization problem giving the solution

Let X and t be given as before as well as t elements $u_1,\ldots,u_t$ from G.

Find a solution $(y_1,y_2,\ldots,y_t)$ *to*

$$y_1y_2 \ldots y_t = 1$$

$$y_1 \in u_1 X \;,\; y_2 \in u_2 X \;,\; \ldots \;,\; y_t \in u_t X \qquad (1)$$

in feasible time.

1.4.3.3. - How the factorization solution solves the fraud problem

The chosen new caracters being

$$b'_1,\ldots,b'_{i_1-1},b'_{i_1+1},\ldots,b'_{i_2-1},\ldots,b'_{i_t-1},b'_{i_t+1},\ldots,b'_k\ ,$$

we define

$$u_1 = b'_{i_t+1}\ b'_{i_t+2}\ \cdots\ b'_k\ s^{-1}\ b'_1\ \cdots\ b'_{i_1-1}\ ,$$

$$u_2 = b'_{i_1+1}\ \cdots\ b'_{i_2-1}\ ,$$

$$u_t = b'_{i_{t-1}+1}\ \cdots\ b'_{i_t-1}\ .$$

Now by solving (1) the t unknown values $x_1,\ldots,x_t$ of the new plaintext are obtained,

$$x_1 = u_1^{-1}y_1\ ,\ x_2 = u_2^{-1}y_2\ ,\ \cdots\ ,\ x_t = u_t^{-1}y_t\ .$$

The result is then straightforward.

1.4.4. - Toward a probabilistic algorithm

We will here introduce the ideas which preside over the settling of a clean probabilistic model used for solving (1).

Let us write $t = 16r$ and

$$y_1\cdots y_r = a_1\ ,\ y_{r+1}\cdots y_{2r} = a_2\ ,\ \cdots\ ,\ y_{t-r+1}\cdots y_t = a_{16}\ .$$

After, put moreover

$$a_1a_2 = b_1\ ,\ a_3a_4 = b_2\ ,\ \cdots\ ,\ a_{15}a_{16} = b_8\ ;$$

$$b_1b_2 = c_1\ ,\ b_3b_4 = c_2\ ,\ \cdots\ ,\ b_7b_8 = c_4\ ;$$

$$c_1c_2 = d_1\ ,\ c_3c_4 = d_2\ .$$

We now consider the chain of subgroups

$$H_0 \le H_1 \le H_2 \le H_3 \le H_4 = G$$

where $H_0 = \{1\}$, H_1 is the group of lower triangular matrices of the form $\begin{bmatrix} 1 & 0 \\ b & 1 \end{bmatrix}$, H_2 is the group of lower triangular matrices of the form $\begin{bmatrix} a & 0 \\ b & 1 \end{bmatrix}$ and H_3 is formed by the matrices of the form $\begin{bmatrix} a & 0 \\ b & d \end{bmatrix}$, $ad \neq 0$. Now the idea is to find a solution for (1) in which $a_1,\ldots,a_{16}$ are matrices of G such that $b_1,\ldots,b_8$ all are in H_3, and $c_1,\ldots,c_4$ are in H_2 and d_1,d_2 in H_1. This is actually a new constraint to the problem. However this permits breaking the algorithms into independent steps each step being easier than finding right away a sequence $y_1,\ldots,y_t$, $y_i \in u_iX$, $i = 1,\ldots,t$ verifying $y_1\ldots y_t = 1$.

Denote u_iX by $\mathbf{U}_i$, $i = 1,\ldots,t$. The general idea is as follows. Given the sets $A_1 \subset \mathbf{U}_1 \times \ldots \times \mathbf{U}_r$ and $A_2 \subset \mathbf{U}_{r+1} \times \ldots \times \mathbf{U}_{2r}$, find a subset $B_1 \subset A_1 \times A_2$ such that the product of the components of each element in B_1 lies in H_3. Construct similar sets $B_2,\ldots,B_8$. If these sets are large enough, we will find a subset C_1 of $B_1 \times B_2$ such that the product of the components of each element in C_1 lies in H_2. We construct similar sets C_2, C_3, C_4. We then find $D_1 \subset C_1 \times C_2$ and $D_2 \subset C_3 \times C_4$ such that the product of the components of each element in D_i, $i = 1, 2$ lies in H_1. Finally we only need one element in D_1 and one in D_2 to form $(y_1,\ldots,y_t)$ such that $y_1y_2\ldots y_t = 1$ and moreover $y_i \in u_iX$, $i = 1,\ldots,t$. This looks feasible since apparently the probability that the product of two elements from G lies in H_3 is close to $1/(p+1) = (p-1)^2p/(p^2-1)(p^2-p)$. Thus an average number of $p+1$ tries should give one success. Similar consideration lead to obtaining products from H_3 in H_2 and so on. Hence it seems that if the size of A_i, $i = 1,\ldots,16$ is large enough, we may succeed within a reasonable number of computations. The problem is the settling of a probabilistic model that enable us to predict the issues of the process. Thatfor, it appeared easier to consider not subsets $A_i, B_i, \ldots$ but sequences $(\phi_1,\ldots,\phi_n)$ and $(\psi_1,\ldots,\psi_n)$ with elements from $\mathbf{U}_1 \times \ldots \times \mathbf{U}_r$ and from $\mathbf{U}_{r+1} \times \ldots \times \mathbf{U}_{2r}$ respectively as a first step of the algorithm and similar sequences for the other steps. This is clearly settled in section two. There, probabilistic properties of those sequences are verified. This allows the computation of the sizes of the sequences needed for a given probability of success of the algorithm. This is done in section 3. A numerical example is dealt with at the end of section 2.

II. - RANDOM VARIABLES WITH UNIFORM PROBABILITY DISTRIBUTIONS RELATED TO GL(2,p)

2.1. - The basic lemma

2.1.1. - Lemma

The mapping $\eta : GL(2,p) \to \mathbb{F}_p \times \mathbb{F}_p^* \times \mathbb{F}_p^* \times (\mathbb{F}_p \cup \{\infty\})$ *defined by*

$\eta((x_{ij})) = (x_{21}x_{11}^{-1}, x_{11}, x_{22}, \infty)$ *when* $x_{12} = 0$ *and*
$((x_{ij})) = (x_{22}x_{12}^{-1}, x_{12}, x_{21}-x_{11}x_{22}x_{12}^{-1}, x_{11}x_{12}^{-1})$ *otherwise, is one-to-one.*

Let us consider the following chain of subgroups of $GL(2,p)$

$$H_0 \le H_1 \le H_2 \le H_3 \le G = H_4$$

where H_0 reduces to the identity; H_1 is the group of lower triangular matrices of the form $\left[\begin{smallmatrix}1 & 0\\ b & 1\end{smallmatrix}\right]$ which is isomorphic to $(\mathbb{F}_p,+)$; H_2 is the group of triangular matrices of the form $\left[\begin{smallmatrix}a & 0\\ b & 1\end{smallmatrix}\right]$, $(b,a) \in \mathbb{F}_p \times \mathbb{F}_p^*$ and H_3 is formed by those matrices of the form $\left[\begin{smallmatrix}a & 0\\ b & d\end{smallmatrix}\right]$, $(b,a,d) \in \mathbb{F}_p \times \mathbb{F}_p^* \times \mathbb{F}_p^*$.

On the other hand let us denote by E_1 the set H_1, by E_2 the set of matrices of the form $\left[\begin{smallmatrix}a & 0\\ 0 & 1\end{smallmatrix}\right]$, $a \in \mathbb{F}_p^*$ and by E_3 the set of matrices of the form $\left[\begin{smallmatrix}1 & 0\\ 0 & d\end{smallmatrix}\right]$, $d \in \mathbb{F}_p^*$. Notice that each of these sets forms a group. Finally E_4 will denote the set of matrices of the form $\left[\begin{smallmatrix}u & 1\\ 1 & 0\end{smallmatrix}\right]$, $u \in \mathbb{F}_p$, together with the identity matrix.

We clearly have $H_1 = E_1$, $H_2 = H_1E_2$, $H_3 = H_2E_3$. Thus $H_3 = E_1E_2E_3$ and here is defined a natural one-to-one mapping from H_3 onto $\mathbb{F}_p \times \mathbb{F}_p^* \times \mathbb{F}_p^*$:

$$\begin{bmatrix} a & 0 \\ b & d \end{bmatrix} \leadsto \begin{bmatrix} 1 & 0 \\ ba^{-1} & 1 \end{bmatrix} \begin{bmatrix} a & 0 \\ 0 & 1 \end{bmatrix} \begin{bmatrix} 1 & 0 \\ 0 & d \end{bmatrix} \leadsto (ba^{-1},a,d)$$

If we now prove that the mapping of $H_3 \times E_4$ into G defined by $(x,y) \to xy$ is surjective, then the lemma will be proved since then

$$(p^2-1)(p^2-p) = |G| = |H_3E_4| \le |H_3 \times E_4| = p(p-1)^2(p+1) .$$

We thus consider any matrix $\begin{bmatrix} x_{11} & x_{12} \\ x_{21} & x_{22} \end{bmatrix}$ of G and we show that it is obtained by the product of a matrix of H_3 by a matrix of E_4. This is obvious when $x_{12} = 0$. Otherwise we have that

$$\begin{bmatrix} x_{11} & x_{12} \\ x_{21} & x_{22} \end{bmatrix} = \begin{bmatrix} x_{12} & 0 \\ x_{22} & x_{21}-x_{11}x_{22}x_{12}^{-1} \end{bmatrix} \begin{bmatrix} x_{11}x_{12}^{-1} & 1 \\ 1 & 0 \end{bmatrix}$$ □

<u>Remarks</u>

The inverse of the mapping given by the lemma is

$$\text{for } d = \infty : (a,b,c,d) \leadsto \begin{bmatrix} b & 0 \\ ab & c \end{bmatrix} ;$$

for $d \in \mathbb{F}_p$: $(a,b,c,d) \rightsquigarrow \begin{bmatrix} db & b \\ dab+c & ab \end{bmatrix}$.

The mapping $x \to (\eta_1(x),\eta_2(x),\eta_3(x),\eta_4(x)) = \eta(x)$ clearly defines four random variables from G onto $\mathbb{F}_p$, $\mathbb{F}_p^*$, $\mathbb{F}_p^*$ and $\mathbb{F}_p \cup \{\infty\}$ respectively for any probability distribution defined on G.

2.1.2. - Right and left mappings

Let us call right mapping the mapping of the lemma. We define similarly a mapping from G onto $\{\mathbb{F}_p \cup \{\infty\}\} \times \mathbb{F}_p^* \times \mathbb{F}_p$ from the factorization of y for $y_{12} \neq 0$:

$$y = \begin{bmatrix} y_{11} & y_{12} \\ y_{21} & y_{22} \end{bmatrix} = \begin{bmatrix} 0 & 1 \\ 1 & y_{22}y_{12}^{-1} \end{bmatrix} \begin{bmatrix} y_{21}-y_{11}y_{22}y_{12}^{-1} & 0 \\ y_{11} & y_{12} \end{bmatrix} .$$

Here we explicitly define the mapping by

$$y \rightsquigarrow (y_{22}y_{12}^{-1}, y_{12}, y_{21}-y_{11}y_{22}y_{12}^{-1}, y_{11}y_{22}^{-1})$$

for $y_{12} \neq 0$ and

$$y \rightsquigarrow (\infty, y_{22}, y_{11}, y_{21}y_{22}^{-1})$$

for $y_{12} = 0$, with the notation

$$y \rightsquigarrow (\theta_4(y),\theta_3(y),\theta_2(y),\theta_1(y)) .$$

Clearly, we have defined new random variables $\theta_1,\theta_2,\theta_3,\theta_4$ having the same properties as $\eta_1,\eta_2,\eta_3,\eta_4$.

The following corollary is straightforward.

Corollary 1

For any couple $(x,y) \in G\times G$, *we have that* $x\,y \in H_3$ *iff* $\eta_4(x) = -\theta_4(y)$. *For* $(x,y) \in H_3 \times H_3$, *we have that* $x\,y \in H_2$ *iff* $\eta_3(x) = (\theta_3(y))^{-1}$. *For* $(x,y) \in H_2 \times H_2$, $x\,y \in H_1$ *iff* $\eta_2(x) = (\theta_2(y))^{-1}$. *For* $(x,y) \in H_1 \times H_1$, $x\,y = I$ *iff* $\eta_1(x) = -\theta_1(y)$.

2.1.3. - The random variables

Let us denote by

$$x \to \eta(x) = (\eta_1(x),\eta_2(x),\eta_3(x),\eta_4(x)) = (x_1,x_2,x_3,x_4)$$

$$\in \mathbb{F}_p \times \mathbb{F}_p^* \times \mathbb{F}_p^* \times (\mathbb{F}_p \cup \{\infty\})$$

$\forall\, x \in G$ the mapping given in the lemma.

Now let us be given a set E with a uniform probability distribution and a random variable ζ from E onto G. Besides, let ξ_1,ξ_2,ξ_3,ξ_4 be random variables from E onto $\mathbb{F}_p$, $\mathbb{F}_p^*$, $\mathbb{F}_p^*$, and $\mathbb{F}_p \cup \{\infty\}$, respectively. We have the

Corollary 2

Let η be such that $\xi_1 = \eta_1 o \zeta$, $\xi_2 = \eta_2 o \zeta$, $\xi_3 = \eta_3 o \zeta$ and $\xi_4 = \eta_4 o \zeta$. Then we have that ζ has a uniform probability distribution iff each of ξ_1,ξ_2,ξ_3,ξ_4, has a uniform probability distribution.

We here just sketch the proof. Let us assume that ζ has a uniform P.D. and then prove that ξ_1 has a uniform P.D.

By definition, for any $x_1 \in \mathbb{F}_p$, $P\{\xi_1=x_1\}$ is the measure of the subset of E mapped by ζ onto G_1, where G_1 is the set of all matrices $z \in G$ such that $\eta_1(z) = x_1$.

Thus $P\{\xi_1=x_1\} = \sum_{z\in G_1} P\{\zeta=x\}$. But $P\{\zeta=z\}$ is a constant, by hypothesis, say $P\{\zeta=z^*\}$ where z^* is any matrix of G, and $P\{\zeta=z^*\} = 1/|G|$. Now by the lemma, $|G_1| = (p-1)^2(p+1)$. Hence $P\{\xi_1=x_1\} = |G_1|/|G| = p^{-1}$.

2.1.4. - Toward applications

Concretely the set E just introduced that we have in view will be a cartesian product $X_1 \times X_2 \times \ldots \times X_r$ of small subsets of G. For example if p is about 10.000, say $p = 10.007$, then $|G|$ is about 10^{16}, we want to make $|X_i| < 100$. The mapping ζ will be defined by $\zeta : (x_1,\ldots,x_r) \to x_1x_2\ldots x_r$. Then there is no hope that ζ could have a uniform P.D. unless r is larger than 8. Actually, we expect that the mapping η of E into G is a random variable with uniform P.D. when r is, say, larger than 20, since requirement 1.3.2. asks for such a uniform P.D.

However since the algorithm in view only deals separately with the random variable $\xi_1, \xi_2, \xi_3, \xi_4$ and if these verify the statistical tests for uniformly distributed random variables, it will be reasonable to consider smaller values of r. Our aim will then be to draw random n-sequences from E with a uniform P.D. and then consider the image under ζ of these n-sequences as random n-sequences from G with a uniform P.D. The probabilistic algorithm will rely on that technique. Observe that such sets E may be used to produce by means of ξ_4 pseudo-random binary sequences for p a Mersenne prime.

2.2. - A recursive probabilistic algorithm

2.2.1. - General considerations on random n-sequences

Given a finite set E with a uniform probability distribution, we shall call random n-sequence drawn from E or random sample of size n an element $(x_1,\ldots,x_n)$ of E^n. This is to imply that all possible samples have the same probability $|E|^{-n}$. If moreover a random variable ζ from E onto $G = H_4$ is defined with a uniform P.D., then ζ and $(x_1,\ldots,x_n)$ yield a random n-sequence $(v_1,\ldots,v_n)$ of G^n. We will now consider two sets E and F, $|E| = |F|$ and two random variables ζ and γ respectively from E and F onto G with the same properties as above. We then define $T \subset E \times F$:

$$T = \{(x,y)/\zeta(x).\gamma(y) \in H_3\}$$

and the random variable $\zeta * \gamma : (x,y) \to \zeta(x)\gamma(y)$ from T onto H_3 is well defined. It has uniform P.D. For,

$$P\{(x,y) \in T\} = |H_3|^2(p+1)/|G|^2 = 1/(p+1) ,$$

by the lemma, and for any $z \in G$, $P\{(x,y)|x\ y = z\} = |G|/|G|^2$.

Thus for $z \in H_3$,

$$P\{\zeta * \gamma = z\} = P\{(x,y)|x\ y = z\}/P\{(x,y) \in T\} = 1/|H_3|.$$

Thus a random n-sequence from $E \times F$ will provide a random n'-sequence from T, but n will be large compared to n' since the average number of drawings is $p+1$ for a single element of T.

The arguments hold here and in the following for passing from H_4 to H_3 as well as for passing from H_i to H_{i-1}, $i < 4$. But this does not provide a good algorithm.

We then need a good algorithm for transforming two random n-sequences, one from E, the other from F into a random r-sequence from T with $r < n$ but with an r not to small compared to n. Afterwards, given two other sets E' and F' defined as E and F, we will obtain a r'-sequence from a set T' defined similarly as T. Denoting by n' the min(r,r'), we then obtain two random n'-sequences, one from T, the other from T'. The algorithm will then proceed recursively. We will show in the next section how to use it for factoring the unity of G and, from there, any element from G into factors to be found in small subsets of G.

The aim of the algorithm that we give here for finding a random r-sequence from T is double. First it needs few operations and secondly, given a probability Π and an integer k, we will be able to determine which n will yield an $r \geq k$ with probability larger than Π.

2.2.2. - The basic theorem

We still denote by G the group GL(2,p) and by H_3 the subgroup of lower triangular matrices.

We will use as before a function $\eta : G \to \{\mathbb{F}_p \cup \{\infty\}\}$, $\eta(x) = x_{11}x_{12}^{-1}$ for $x_{12} \neq 0$ and $\eta(x) = \infty$ for $x_{12} = 0$; also $\theta : G \to \{\mathbb{F}_p \cup \{\infty\}\}$, $\theta(x) = x_{22}x_{12}^{-1}$ for $x_{12} \neq 0$ and $\theta(x) = \infty$ for $x_{12} = 0$.

Then our algorithm will rely on the

Theorem 1

Let $(w_1,\dots,w_n)$ *be a random n-sequence from* $G = GL(2,p)$. *Reorder it as* $(u_1,\dots,u_n)$ *so that* $\theta(u_i) \leq \theta(u_j)$ *for* $i < j$. *Now let* $(v_1,\dots,v_n)$ *be any given n-sequence from* G *verifying* $\eta(v_i) = -\theta(u_i)$, $i = 1,\dots,n$. *Then* $(v_1u_1,\dots,v_nu_n)$ *is a random n-sequence from* H_3.

We first observe that $v_iu_i \in H_3$, $i = 1,\dots,n$. Denoting v_i by x and u_i by y, we have indeed $x_{11}y_{12} + x_{12}y_{22} = 0$ from $\eta(v_i) = -\theta(u_i)$ (Corollary 1). Now, by the lemma, we may write, using the left mapping, $(w_1,\dots,w_n) = ((x_1,w_1'),\dots,(x_n,w_n'))$ where $(x_1,\dots,x_n)$ is first drawn from $\mathbb{F}_p \cup \{\infty\}$ and then

$(w_1',\dots,w_n')$ is drawn independently from H_3. We here have $x_i = -\theta(w_i)$. Reordering $(x_1,\dots,x_n)$ as $(y_1,\dots,y_n)$ so that $y_i \leq y_j$ for $i < j$ defines once for all a permutation on the indices. That permutation yields a permutation of H_3^n which preserves its uniform probability distribution. Hence $(u_1,\dots,u_n) = ((y_1,u_1'),\dots,(y_n,u_n'))$ where $y_i = -\theta(u_i)$ and where $(u_1',\dots,u_n')$ is a random n-sequence from H_3. Now $(v_1,\dots,v_n)$ is as well $((v_1',-y_1),\dots,(v_n',-y_n))$ where $(v_1',\dots,v_n')$ is any given fixed n-sequence from H_3. Here we used the right mapping of the lemma. Here again multiplying componentwize by $(v_1',\dots,v_n')$ every element of H_3^n defines a permutation of H_3^n which preserves its uniform probability distribution. But from the properties of right and left mapping of the lemma, we have that $(v_1u_1,\dots,v_nu_n) = (v_1'u_1',\dots,v_n'u_n')$ which is as just shown, a random n-sequence from H_3. □

2.2.3. - The algorithm

The essentiel problem to be solved is the following. Given two sets E and F with $|E| = |F|$ as in 2.2.1., draw two random n-sequences, one from E, the other from F and then determine the two corresponding random n-sequence $(v_1,\dots,v_n)$ and $(w_1,\dots,w_n)$ from $G = H_4$. The purpose is from there to construct one r-sequence from H_3 together with the corresponding r-sequence from T with r "not to small" compared to n. The algorithm will be repeated, replacing H_4 by H_3 and H_3 by H_2, and so on. We will have to fix n large enough to finally obtain a 1-sequence in H_0 which reduces to the identity matrix. Assume that the value of n is known. After determining the two random n-sequences $(v_1,\dots,v_n)$ and $(w_1,\dots,w_n)$ from $G = H_4$, then the following steps of the algorithm are :

2. Reorder $(w_1,\dots,w_n)$ as $(u_1,\dots,u_n)$ by sorting $\{w_i\}_{i\in[n]}$ according to the values $-\theta(w_i)$. (See proof of theorem 1).

3. Reorder $(v_1,\dots,v_n)$ according to the values of $\eta(v_i)$

4. Determine the set

$$S = \{\eta(v_i) \mid i \in [n]\} \cap \{-\theta(w_i) \mid i \in [n]\}$$

(This need less than $2n$ comparisons help to 2 and 3).

5. Determine the r-subsequence $(u_1^*,\dots,u_r^*)$ of $(u_1,\dots,u_n)$ with $-\theta(u_i^*) \in S$, $i = 1,\dots,r$.

6. Construct an r-sequence $(t_1,\dots,t_r)$ with terms taken from $\{v_1,\dots,v_n\}$ such

that $\eta(t_i) = -\theta(u_i^*)$, $i = 1,\dots,r$.

7. Compute the r-sequence $(t_1u_1^*,\dots,t_ru_r^*)$ from H_3 and keep in memory the corresponding sequence from $T \subset E \times F$.

In section 2.3 we deal with numerical values. The numbers n, r and $s = |S|$ that we are concerned here with will be denoted n_4, r_4 and s_4 in the general algorithm. We start with $r_1 = 1$ and decide of the values of n_1, s_1. Then $r_2=n_1$ and we decide of the values of n_2, s_2. Finally we have $n_3 = r_4$ and the value $n = n_4$ of the above algorithm will be determined.

Notice that we have to replace $-\theta(u_i^!)$ by $(\theta(u_i^!))^{-1}$ when computing an r-sequence from H_2 or from H_1.

2.3. - Numerical values

2.3.1. - First step

The first step consists in determining the size of the random n_1-sequences from H_1 such that the product of one selected term in the first by one term in the second sequence gives the identity matrix.

We know that H_1 may be identified with $\mathbb{F}_p$. We take $p = 10,007$. We thus first have a mapping $a : [n_1] \to [m]$ with $m=10,007$ and with $|\mathrm{Im}a| \geq s_1$. Then mapping $b : [n_1] \to [m]$ defining the second n_1-sequence from H_1 may be considered as a sequence of n_1 independant Bernoulli trials, each with a probability of success s_1/m. Since we need at least one success with probability $1-10^{-3}$, we necessarily fix s_1 and n_1 such that

$$(1-s_1/m)^{n_1} < 10^{-3} .$$

We have a degree of freedom in that choice but we jointly need that the probability that $|\mathrm{Im}\, a| \geq s_1$ be at least $1-10^{-3}$. For

$$n_1 = 297 \quad , \quad s_1 = 285 ,$$

we have that

$$(1-s_1/m)^{n_1} < 1.8810^{-4} .$$

On the other hand we compute an upper bound for

$$\varepsilon_{s_1,n_1} = m^{-n_1} \sum_{1\le i< s_1} (m)_i\ S(n,i) ,$$

where $S(n,i)$ is a Stirling number of the second kind, (see the Appendix).

We have from property 1, in the Appendix that

$$\varepsilon_{s_1,n_1} = (m-1)_{s_1-1} \sum_{i\ge 0} m^{-n_1-i}\ S(n_1+i,s_1-1) .$$

We use the upper bound $S(n,s) < s^n/s!$ which is not tight but quite satisfactory for the present purpose. Notice that

$$\lim_{n\to\infty} S(n,s)/s^n = 1/s! .$$

Thus using that upper bound is convenient in the application in view except in this first step where we compute exactly

$$m^{-n_1} \sum_{s_1\le i\le n_1} (m)_i\ S(n,i) = 1-\varepsilon_{s_1,n_1} = 0.999572 \ldots ,$$

for $n_1 = 297$ and $s_1 = 285$.

Notice that without the constraint that n_1 should be larger than s_1 from the condition $\varepsilon_{s_1,n_1} < 10^{-3}$, we would have had the only condition

$$-n_1 s_1/m < \log 10^{-3}$$

which gives for $n_1 = s_1$,

$$n_1 = s_1 = 263 \ ; \ (1-263/10{,}007)^{263} < 9.810^{-4} .$$

From this we observe that the Bernoulli constraint prevails. Next, we use $s!\ S(n,s) < s^n$ to bound $\varepsilon_{s,n}$ by a geometric series of which the sum is

$$\binom{m-1}{s-1} \left(\frac{s-1}{m}\right)^n \frac{m}{m-s+1} .$$

The reader is invited to see the Appendix for further details.

2.3.2. - Second step

The scheme is the same as for the first step except that we here have to consider two n_2-sequences from $\mathbb{F}_{p^*}$. Then $m = 10{,}006$. The second n_2-sequence consist in n_2 bernoulli trials and we must have $r_2 = 297$ success with probability larger than $1-10^{-3}$. Thus

$$\sum_{0\leq j<r_2} \binom{n_2}{j} p_{s_2}^j \, q_{s_2}^{n-j} < 10^{-3}$$

with $p_{s_2} = s_2/m$, where the probability that the first n_2-sequence from $\mathbb{F}_p^*$ gives at least s_2 distinct terms must be larger than $1-10^{-3}$. The binomial tail is estimated by using the normal distribution. We have that

$$\mathcal{N}(3.11) = 0.9991 \ldots ;$$

we have that

$$P\{n_2 p_{s_2} + 3.11\sqrt{n_2 p_{s_2} q_{s_2}} \geq r_2\} = \mathcal{N}(3.11)$$

See for example W. FELLER [2].

We find out that for $n_2 = s_2$ the value $n_2 = s_2 = 1{,}870$ gives the expected inequality. We now have to spread the values of n_2 and s_2 in order to obtain $\varepsilon_{s_2,n_2} < 10^{-3}$ together with the previous inequality.

We find out that $n_2 = 2{,}287$ and $s_2 = 1{,}534$ still satisfy the previous inequality and moreover $\varepsilon_{s_2,n_2} < 10^{-3}$ for $n_2 = 2{,}286$ and $s_2 = 1{,}534$. We keep $n_2 = 2{,}287$.

2.3.3. - Third step

Here again $m = 10{,}006$. We put $r_3 = 2{,}287$ and we try equal values n_3, s_3 for the Bernoulli tail. We obtain $n_3 = s_3 = 4{,}897$. By spreading those values, we have that $n_3 = 6{,}562$ with $s_3 = 3{,}673$ maintain the required inequality. On the other hand the upper bound for ε_{s_3,n_3} makes us sure that $\varepsilon_{s_3,n_3} < 10^{-3}$ for $n_3 = 6{,}564$ and $s_3 = 3{,}673$. We thus make $n_3 = 6{,}564$.

2.3.4. - Last step

Here $m = 10{,}008$ and $r_4 = 6{,}564$. We obtain $n_4 = s_4 = 8{,}172$ when asking for equal values in the inequality for the binomial tail. Now spreading the values of n_4 and s_4, we get $n_4 = 11{,}952$ and $s_4 = 5{,}638$ in order to obtain 6,564 success by n_4 trials with probability p_{s_4}. Finally the upper bound used shows that we should draw at least 11,940 elements with replacement from a set of 10,008 elements in order to obtain 5,638 distincts elements with probability $1-10^{-3}$.

2.3.5. - Final decision

The final result is that we should start with $n_4 = 11,952$ for a global probability of success at least $(1-.002)^{15} > 97\%$. We start again if we failed.

2.4. - Conclusion

Coming back to the notations of 1.4.4, we need 16 n_4-sequences to start with, one from each $\mathbf{U}_{ir+1} \times \mathbf{U}_{ir+2} \times \ldots \times \mathbf{U}_{(i+1)r}$, $i = 0,\ldots,15$, with $n_4 = 11,952$.

Let us now consider the value of r which is to be fixed in order that the basic probabilistic hypothesis be verified. The size of each set from which the elements of the n_4-sequences are drawn is 64^r. Each of those sets is a set E as considered in 2.1.3. Since the size of G is about 10^{16}, it is hopeless that ζ have a uniform probability distribution unless $r \geq 9$. However, all we need is that each random variable ξ_i, $i = 1,\ldots,4$ behaves as a uniformly distributed random variable. Indeed at each step of the algorithm, we only deal with one of the random variables ξ_i. Thus it is very likely that we don't actually need that ζ have a uniform probability distribution. For $r = 3$, for example, ξ_4 is a mapping from a set of $64^3 = 262,144$ elements into a set of $p+1 = 10,008$ elements and it is very likely that ξ_4 will behave as a uniformly distributed random variable and that the algorithm will proceed as expected even for that small fixed value of r. Indeed, in the second step of the algorithm, we will consider an n_3-sequence with $n_3 = r_4 = 6,564$ from $\mathbf{U}_1 \times \mathbf{U}_2 \times \mathbf{U}_3 \times \mathbf{U}_4 \times \mathbf{U}_5 \times \mathbf{U}_6$. The n_3-sequence $(t_1u_1^*, t_2u_2^*, \ldots, t_{n_3}u_{n_3}^*)$ is thus drawn from $H_3 \cap \mathbf{U}_1 \mathbf{U}_2 \ldots \mathbf{U}_6$ of which the size may be roughly estimated as $64^6/10^4 \simeq 6.810^6$. For that reason, one may expect that the further projection η_3 which maps that set into $p-1 = 10,006$ elements will behave as a uniformly distributed random variable although $t = 16r$ with $r = 3$ is theoretically to small a value for t.

III. - APPENDIX

3.1. - Probabilities related to drawing two random n-sequences from an m-set

3.1.1. - The probability $P(r,n,m)$

The m-set, identified with the set $[m]$ of the first m integers, has a uniform probability distribution, briefly U.P.D. Such n-sequences may be viewed as mappings from the n-set $[n]$ into the m-set $[m]$. Let a and b be two such mappings. In view of the algorithm introduced in section 2.2 we have to consider the probability $P(r,n,m)$ that less than r elements of $[n]$ are mapped by b into the image of $[n]$ by a. Thus

$$P(r,n,m) = P\{|b^{-1}(\text{Imb} \cap \text{Ima})| < r\} .$$

This is the probability of failure at each probabilistic step of the algorithm (there are 15 such steps) for various values of the parameters r, n, m. In the numerical example dealt with in section 2.3, we asked that at each step $P(r,n,m)$ should be smaller than 2.10^{-3}.

3.1.2. - The value of $P(r,n,m)$

Property 1

$$P(r,n,m) = m^{-n} \sum_{1 \le i \le n} (m)_i \, S(n,i) \sum_{0 \le j < r} \binom{n}{j} p_i^j q_i^{n-j} \qquad (1)$$

where $p_i = i/m$, $q_i = 1-i/m$, $(m)_i = m(m-1)\dots(m-i+1)$ *and* $S(n,i)$ *is a Stirling number of the second kind.*

Proof

$S(n,i)$ is the number of partitions of $[n]$ into i classes and consequently, $(m)_i\, S(n,i)$ is the number of mappings $a : [n] \to [m]$ such that $|\text{Ima}| = i$. To each such mapping corresponds a probability

$$\sum_{0 \le j < r} \binom{n}{j} p_i^j q_i^{n-j}$$

that a mapping $b : [n] \to [m]$ maps less than r elements in the set Ima. □

3.1.3. - Toward a computable upper bound for $P(r,n,m)$

In section 2.3 we use a dynamic programming argument to decide of the size n_4 of the first n-sequence to be drawn. At each step we could use expression (1) with r and m given and compute n large enough for $P(r,n,m)$ to be smaller than 2.10^{-3}.

However m is close to 10^4 and n may be as large. Thus computing (1) by first tabulating $S(n,i)$ and $\binom{n}{j}$ using recurrence relations looks beyond reach. Our purpose will thus be to set up an upper bound for the probability of failure $P(r,n,m)$. Even if the bound is not tight, so long as it is easily computed and provides reasonable values for n at each step of the computation as in section 2.3 then the scheme of J. BOSSET will be shown to be breakable. The first step for obtaining such an upper bound is

Property 2

Let $1 \le s \le n$ *and* $\varepsilon_s = m^{-n} \sum_{1 \le i < s} (m)_i \, S(n,i)$. *Then we have that*

$$P(r,n,m) < \varepsilon_s + \sum_{0 \le j < r} \binom{n}{j} p_s^j \, q_s^{n-j}. \qquad (2)$$

Proof

On the one hand, we have that

$$\sum_{0 \le j < r} \binom{n}{j} p_i^j \, q_i^{n-j}$$

is upper bounded by 1 for any i and in particular for $i < s$. On the other hand $m^{-n} \sum_{s \le i < n} (m)_i \, S(n,i)$ is $1-\varepsilon_s$.

Consequently all is left to be proved is that if we denote by $f(x,r)$ the expression

$$\sum_{0 \le j < r} \binom{n}{j} p_x^j \, q_x^{n-j} ,$$

we have that $f(s,r) \ge f(i,r)$ for every $i > s$.

This actually means that the probability to have few success (less than r) in running n Bernoulli trials is larger when the probability p_s of success is smaller.

Denote by $g_{s,i}(r+j)$ the difference $f(s,r+j) - f(i,r+j)$, $j = 0,\dots,n-r+1$. We then have to prove that $g_{s,i}(r)$ is nonnegative for $i > 1$. We first have $g_{s,i}(1) = q_s^n - q_i^n > 0$ and $g_{s,i}(n+1) = 0$.

Now we have that

$$g_{s,i}(r+1) = f(s,r+1) - f(i,r+1) = f(s,r) - f(i,r) + \Delta_{r+1} ,$$

where

$$\Delta_{r+1} = \binom{n}{r} ((p_s q_s^{-1})^r q_s^n - (p_i q_i^{-1})^r q_i^n) ,$$

$$\Delta_{r+1} = r^{-1}(n-r+1)((p_s q_s^{-1})\Delta_r + \binom{n}{r-1}(p_s q_s^{-1} - p_i q_i^{-1})(p_i q_i^{-1})^{r-1} q_i^n) .$$

From the fact that $p_s q_s^{-1} - p_i q_i^{-1} < 0$, we see that $\Delta_{r+j} < 0$ implies $\Delta_{r+j+1} < 0$, $j \le n-r$.

This means that if $g_{s,i}(r+j)$ eventually starts decreasing, then it goes on decreasing down to $g_{s,i}(n+1)$ which is zero. Hence we have that $g_{s,i}(r)$ is nonnegative. □

3.2. - <u>A first step to numerical computation</u>

We are not finished with setting up an upper bound to $P(r,n,m)$ since the number ε_s should be itself again upper bounded. This will be the purpose of another section in this Appendix. What we can do right now is obtaining numerical values for n and s in order that $f(s,n) < 10^{-3}$. We actually compute, for given r, values of n and s in order that the inequality

$$np_s - 3.11\sqrt{np_s q_s} \ge r \qquad (3)$$

holds.

This means that the probability that the number of success **be smaller** than $np_s - 3.11\sqrt{np_s q_s}$ will be close to 0.0009. Thus the probability that the number of success is $r-1$ or less will be still smaller.

Inequality (3) is given by DE MOIVRE-LAPLACE limit theorem (see for example W. FELLER [2]).

3.3. - An easily computable upper bound for $\varepsilon_s = m^{-n} \sum_{1\le i\le s} (m)_i \, S(n,i)$

3.3.1. - The direct computation

First, if n is not too large, a direct computation is feasible. We first compute by recurrence

$$S(n,k) = S(n-1,k-1) + k\, S(n-1,k) \tag{3}$$

with

$$S(n,1) = 1 \quad \text{and} \quad S(n,n) = 1 \quad \text{for all } n.$$

Since we need an upper bound for ε_s, we may use for $m^{-n}(m)_i$ the upper approximation

$$m^{-n}(m)_i < (m/(m-i))^{m-i+.5} \exp(-i+1/12m)\, m^{-n+i} \tag{4}$$

given by W. FELLER [2].

But since s is expected not to be much smaller than n, the computing by recurrence of $S(n,s),\dots,S(n,n)$ needs few operations and we better compute $1-\varepsilon_s$ using

$$m^{-n}(m)_i > (m/(m-i))^{m-i+.5} \exp(-i-1/12))m^{-n+i} . \tag{5}$$

3.3.2. - An infinite sum for P(r,n,m)

Property 3

$$m^{-n} \sum_{1\le i<s} (m)_i \, S(n,i) = (m-1)_{s-1} \sum_{i\ge 0} m^{-n-i} \, S(n+i,s-1) . \tag{6}$$

Proof

We first prove that for every $k < m$, we have that

$$(m-1)_k \sum_{i\ge 0} m^{-k-i} \, S(k+i,k) = 1 . \tag{7}$$

Equality (7) is obtained right away by substituting m^{-1} to u in the generating series

$$u^k(1-u)^{-1}(1-2u)^{-1} \dots (1-ku)^{-1} = \sum_{i\geq 0} S(k+i,k)u^{k+i}$$

which is found for example in L. COMTET [3]. Now equality (6) will follow from proving that

$$m^{-n} \sum_{s\leq i\leq n} (m)_i \, S(n,i) = (m-1)_{s-1} \sum_{0\leq j\leq n-s} m^{-s-j+1} S(s+j-1,s-1) \; . \tag{8}$$

The L.H.S. is the probability to obtain at least s distinct elements when drawing n elements from an m-set. That event may be considered the union of $n-s+1$ disjoint events. The jth event, $j = 0,\dots,n-s$, is to obtain for the first time s distinct elements at the $(s+j)^{th}$ drawing. The probability of the jth event is the product of the probability of having drawn exactly $s-1$ distinct elements at the $(s+j-1)^{th}$ drawing by the probability of drawing an element from the complementary set of size $m-s+1$ at the $(s+j)^{th}$ drawing. Hence the probability of the jth event is

$$m^{-s-j+1}(m)_{s-1} \, S(s+j-1,s-1).(m-s+1).m^{-1} \; .$$

This completes the proof of equality (8) and consequently of equality (6). □

3.3.3. - Applying Property 3 to numerical computation

It is well known that

$$\lim_{n\to\infty} s! \, S(n,s)s^{-n} = 1 \tag{9}$$

and moreover $S(n,s) < s^n/s!$

See L. COMTET [3] page 293, exercice 9.

From

$$s^n = \sum_{1\leq i\leq s} (s)_i \, S(n,i) \tag{10}$$

(9) may be obtained by observing that (10) implies

$$1 > s! \, S(n,s)s^{-n} > 1-(s^{-1}(s-1))^n\binom{s}{s-1}-(s^{-1}(s-2))^n\binom{s}{s-2} - \dots - s^{-n} s, \tag{11}$$

$$1 > s! \, S(n,s)s^{-n} > 1-((\exp(-n/s)+1)^s-1-\exp(-n)) \; .$$

From this and Property 3, then ε_s is upper bounded by the geometric series

$$\binom{m-1}{s-1} m^{-n} (s-1)^n \sum_{i\geq 0} ((s-1)m^{-1})^i = \binom{m-1}{s-1} ((s-1)m^{-1})^n m(m-s+1)^{-1} \tag{12}$$

And (9) shows that when i becomes large, each term of the geometric series becomes a tight upper bound for the corresponding term of the R.H.S. of (6).

For a given probability of failure e^{-t} and given s and m, we are now able to compute an n such that the L.H.S. of (6) is upper bounded by e^{-t} :

$$n > (\log(m^{-1}(s-1)))^{-1}(-t+\log(1-m^{-1}(s-1)) - \sum_{0\leq i<s-1} (\log(m-i-1)-\log(i+1))). \tag{13}$$

Another upper bound for $S(n,s)$ permitting to bound (6) by a geometric series appears efficient when $n-s$ is expected to be small and when m is large. We actually need for applying that bound that $c < m$, with $c = s(s+1)/2$. If $n-s$ is small, the upper bound c^{n-s} to $S(n,s)$ which results from Theorem D, page 207 of L. COMTET [3], may lead to satisfactory results, as we here show on a numerical example. However this does not apply to the numerical example of section 2.3 since there c is larger than m.

The upper bound here is

$$c^{n-s+1}\, m^{-(n-s+1)}\, m^{-s} (m)_s\, m(m-c)^{-1} , \tag{14}$$

so that for a given probability of failure e^{-t} and given s and m, the L.H.S. of (6) is upper bounded by e^{-t} as soon as

$$n > (-t+(s-1)(\log c - \log m) - c(m-c)^{-1} (1-c(2m)^{-1}) \tag{15}$$

$$-(m-s+0.5)s(m-s)^{-1}(1-s(2m)^{-1}) + s-(12m)^{-1})/\log(c\, m^{-1}) .$$

We use in that formula the inequalities

$$(m(m-s)^{-1})^{m-s+0.5} \exp(-s+(12m)^{-1}) > (m)_s\, m^{-s} \quad \text{as well as}$$

$$-\log(1-cm^{-1}) < c(m-c)^{-1} (1-c(2m)^{-1}) .$$

Example

For $m = 10^8$, $s = 2{,}300$, $e^{-t} = 10^{-6}$ the R.H.S. of (15) is worth 2,302.803. We thus fix $n = 2{,}303$. On the other hand, the R.H.S. of (13) gives 2,515.096.

3.4. - Computing the average size of Ima for a random mapping $a : [n] \to [m]$

Taking the nth derivative in t of the generating series

$$(e^t + z-1)^m = \sum_{n\geq 0} \sum_{0\leq s\leq m} (m)_s \, S(n,s) \, z^{m-s} \, t^n/n! \; ,$$

we obtain

$$\sum_{0\leq j\leq m} \binom{m}{j} (z-1)^{m-j} \, j^n \, e^{jt} \; ,$$

which gives for $t = 0$

$$f(z) = \sum_{0\leq j\leq m} \binom{m}{j} (z-1)^{m-j} \, j^n = \sum_{0\leq s\leq m} (m)_s \, S(n,s) \, z^{m-s} \; .$$

The average size $\bar{s}$ of Ima for a random mapping $a : [n] \to [m]$ is thus

$$\bar{s} = m - m^{-n} f'(z)\big|_{z=1} = m-(m-1)^n \, m^{-n+1} \; .$$

Thus conversely, if we want to know to which n for a given value of m will correspond an average size $\bar{s}$ we have that

$$n = \log(1-\bar{s}\, m^{-1})/\log(1-m^{-1}) \; . \tag{16}$$

Example

In 2.3.3. we found out that for $m = 10{,}006$ we have a probability larger than $1-10^{-3}$ that $|\mathrm{Ima}| \geq s_3 = 3{,}673$ when the size of n is $6{,}564$. Here, for an average size of $3{,}673$, we find by (16) $n = 4{,}576.624$.

IV. - SOME COMBINATORIAL IDENTITIES RAISED BY INVESTIGATING UPPER BOUND

Relation (8) of section 3.3.2. may write

$$\sum_{s\leq i\leq n} (m)_i \, S(n,i) = (m)_{s+1} \sum_{0\leq j<n-s} m^{n-s-j-1} \, S(s+j,s) \; .$$

This is a polynomial relation of degree n in m and since it holdsfor infinitely many values of m, we may replace m by the indeterminate u for obtaining the polynomial identity of degree $n-s-1$.

$$\sum_{0\leq i<n-s} (u-s-1)_i \, S(n,i+s+1) = \sum_{0\leq i<n-s} u^i \, S(n-i-1,s) \; , \tag{1}$$

where $(u-s-1)_0 = 1$.

We observe that the R.H.S. is a polynomial of $\mathbb{Z}[u]$ expressed in the basis $(u^i)_{i\in\mathbb{N}}$ while the L.H.S. expresses the same polynomial in the basis $((u-s-1)_i)_{i\in\mathbb{N}}$. Since the derivative operator D maps the basis $(u^i/i!)_{i\in\mathbb{N}}$ of $\mathbb{Q}[u]$ onto itself and the difference operator Δ maps onto itself the basis $((u-s-1)_i/i!)_{i\in\mathbb{N}}$ of $\mathbb{Q}[u]$, the use of those operators permit deriving relations among the Stirling numbers of the second kind. We will first observe how (1) entails easily well known recurrence relations.

Substituting s+1 to u and then changing s+1 to k yields

$$S(n,k) = \sum_{0\le i\le n-k} k^i \, S(n-i-1,k-1) \,. \tag{2}$$

Which is quoted as vertical recurrence relation in L. COMTET [3], chap. 5, Th. B [3d]. The same author gives in Th C. [3e] an horizontal relation.

$$S(n,s) = \sum_{0\le i\le n-s} (-1)^i <s+1>_i \, S(n+1,s+i+1) \,, \tag{3}$$

where $<x>_i$ writes $x(x+1)\ldots(x+i-1)$.

This follows from putting $u = 0$ in (1). We now show what can be derived by applying t times the Δ operator to both sides of (1) and then substituting s+1 to u. We recall that $\Delta f(x) = (E-I)f(x) = f(x+1)-f(x)$.

$$t!\, S(n,s+t+1) = \sum_{t\le i<n-s} \Delta^t u^i\big|_{u=s+1} \, S(n-i-1,s) \,.$$

Now

$$\Delta^t u^i = (E-I)^t u^i \,.$$

$$\Delta^t u^i\big|_{u=s+1} = \sum_{0\le j\le t} \binom{t}{j} (s+1+j)^i (-1)^{t-j}$$

$$= \sum_{0\le k\le i-t} (s+1)^k \binom{i}{k} \sum_{0\le j\le t} j^{i-k} \binom{t}{j} (-1)^{t-j} \,.$$

The last factor in that sum appears to be $t!\, S(i-k,t)$. We thus have that

$$\Delta^t u^i\big|_{u=s+1} = t! \sum_{t\le j\le i} S(j,t)(s+1)^{i-j} \binom{i}{j} \,. \tag{4}$$

Actually, the range of j may remain undefined in (4).

We then obtain the

<u>Theorem</u>

$$S(n,s+t) = \sum_{t-1\le i<n-s} \ \sum_{t-1\le j\le i} \binom{i}{j} S(j,t-1)(s+1)^{i-j} S(n-i-1,s) . \qquad (5)$$

The theorem may be considered a generalization of (2) since from the recurrence

$$S(n,k) = k\, S(n-1,k) + S(n-1,k-1) , \qquad (6)$$

we may write $S(j,0) = 0$ for $j > 0$ and $S(0,0) = 1$. On the other hand the ranges of i and j may be left undefined in (5), then the R.H.S. being a convolution product and since

$$\sum_{n\ge 0} S(n,s+t)u^n = u^t(1-(s+1)u)^{-1} \ldots (1-(s+t)u)^{-1} u^s(1-u)^{-1} \ldots (1-s\,u)^{-1} ,$$

we have the

<u>Corollary 1</u>

$$u^t(1-(s+1)u)^{-1} \ldots (1-(s+t)u)^{-1} = \sum_{n\ge 0} \ \sum_{t-1\le j\le n} \binom{n}{j} S(j,t-1)(s+1)^{n-j}u^{n+1} . \qquad (7)$$

We observe that since

$$(1-(s+2)u)^{-1} - (1-(s+1)u)^{-1} = u(1-(s+2)u)^{-1}\,(1-(s+1)u)^{-1} , \qquad (8)$$

then generalizing (2) by (5) with $t = 2$ remains easy. It gives

$$S(n,s+2) = \sum_{1\le i<n-s} ((s+2)^i - (s+1)^i)\, S(n-i-1,s) . \qquad (9)$$

Relations (6) and (8) show how the sequences $(S(n,i) \bmod 2)$, $i = 1,\ldots,n$ behave.

Also Corollary 1 gives for $s = 0$

$$S(n+1,t) = \sum_{t-1\le j\le n} \binom{n}{j} S(j,t-1) \qquad (10)$$

which is relation [3c] in Theorem B already quoted from L. COMTET [3]. From Corollary 1 we also obtain

Corollary 2

$$S(n+1,t) = \sum_{t-1 \le j \le n} (-1)^{t-j+1} \, t^{n-j} \binom{n}{j} S(j,t-1) \qquad (11)$$

Proof

From the generating series for $S(n,t)$ we know that

$$S(n,t) = \sum_{i_1+\dots+i_t=n-t} 1^{i_1} \dots t^{i_t} .$$

Consequently, putting $s = -t-1$ in (7), we have that

$$u^t(1+t\,u)^{-1} \dots (1+u)^{-1} = \sum_{n \ge 0} (-1)^{n-t} S(n,t)u^n \qquad \square$$

REFERENCES

[1] J. BOSSET : "Contre les risques d'altération, un système de certification des informations", 01 Informatique n° 107, Février 1977.

[2] W. FELLER : "An introduction to probability theory and its applications", Wiley, 1968.

[3] L. COMTET : "Advanced combinatoris", D. Reidel, 1974.

Manipulation of Recurrence Relations in Computer Algebra

Jacques CALMET
LIFIA, Grenoble, France

1 Introduction

This talk is devoted to an attempt to manipulate recurrence relations (RR) with computer algebra systems. In studying this problem we have two goals. The first one is to design a computational tool by using a bottom to top approach. The second and most important one is that some of our motivations are closely connected to another project aiming at designing a new system for algebraic and symbolic manipulations [1]. Our goal is to design a module having the following capabilities and features.

1. Identification of recurrence relations: given as input an ordered set of objects determine if they are linked by a RR.
2. Solving (some) recurrence relations.
3. Manipulation of special functions.
4. Design of databases of properties and values of special functions and recurrence relations.
5. Impact of this approach on the design of a new system.

These goals are almost direct consequences of our motivations.

1. In some physical problems based on the calculation of moments as a perturbative expansion, it is guessed that the different approximations obey a RR. It is obvious to check this feature at low orders of the series but impossible for high orders. In fact, the proof is impossible either by hand calculation or by using both intuition and a computer algebra system. The problem in this case is to identify the RR satisfied by a set of given objects.
2. Special functions and recurrence relations have been studied extensively by mathematicians in the past. Nowadays there is a renewed interest in their study, mainly because of the advent of computers and their numerical capabilities. Indeed, engineers and physicists handle special functions almost daily in their calculations. Most often they appear as simple objects such as Legendre polynomials but, sometimes as more elaborated ones such as confluent hypergeometric series. RR are used also in modelling some dynamical systems for example. It may be thus surprising to notice that computer algebra systems are unable to manipulate efficiently these objects. The main exceptions are MACSYMA [2-5], SAC1 [6] and SMP which offer some limited capabilities in that respect. There is thus a gap which must be filled in.
3. RR and special functions have obvious links. It looks therefore feasible to consider a common approach to introduce them in computer algebra.

4. Once it has been decided to design a module of procedures manipulating special functions, it becomes obvious that a data base of their values and/or properties is useful. For instance, there is no reason to recalculate from scratch the value of an orthogonal polynomial for a same argument every time it is needed. This is a waste of computing time. This remark implies that the module has so-called apprentice capabilities and that the system is not only able to compute but must also represents some selected aspects of mathematical knowledge.
5. For RR linking simple objects such as orthogonal polynomials, few theoretical problems have to be solved at the implementation level. But, when dealing with hypergeometric series some problems arise. One of them is the choice of a canonical representation. Indeed, there are so many RR satisfied by these objects that one has to select a basis to represent them. This can be studied either from a mathematical point of view in the spirit of Gosper [3] or from a computer science point of view which consists in investigating the use of term rewriting systems for simplification purposes.
6. Another feature of this project is that no global constructive algebraic method is available. This means that it is practically impossible to cover all of the many occurences of RR known to exist. This implies also that heuristic methods have to be implemented. This is going against the present trend in computer algebra which rejects non-constructive methods from systems. One of the earlier decisions was to concentrate on the implementation of simple methods first and to avoid all aspects presenting purely mathematical difficulties.
7. Points 2 to 6 show that we have at hand a suitable training ground to test some of the capabilities we want to include in a new type of computer algebra system [1]. So far the systems selected for implementation are MACSYMA at the origin, and since it is available REDUCE 3. Because of this motivation the code produced is not prepared for release.

This talk is mainly concerned with the identification of RR. This part was initiated with I. Cohen. A preliminary and incomplete version was presented at RYMSAC II [7]. To solve RR is briefly mentioned. This latter part is a collaboration with Y. Ahronovitz.

2 Recognizing Recurrence Relations

The problem we are interested in is the following one: when a set of objects is given as input, find the RR, if any, these objects satisfy. We assume that this set is ordered. The first elements thus determine the limit conditions for the RR. Depending on the kind of RR sought for enough term must be given in the input. If not, the procedure is not completed.
The first question one must ask is: does it exist a method capable of deciding whether or not a RR holds? The answer is yes. It is the ϵ-algorithm of Brezinski [8]. In addition it gives the order of the RR. A description of this fairly simple algorithm is given in [8]. It is a direct application of Padé approximants and Hankel determinants. This is a good example of a method designed for numerical analysis and also implementable in computer algebra.

As mentioned previously it is not possible to set up a program capable of identifying all RR obeyed by all possible types of objects. We thus define different classes of RR by the type of their arguments. Then, within each class we select those (depending on their order and on their structure) we want to handle.
The ϵ-algorithm is used to select some of the RR which have to be considered.

2.1 Numbers

It is an obvious statement to write that sequences of numbers play an important role in number theory. Since we think that computer algebra systems must have capabilities in this field, we have included this type of RR. Sloane's book [9] displays a very large set of examples. Our immediate goal is to handle all of the sequences in [9] which result from RR.
When looking for a RR for a sequence of numbers the standard approach is to use the method of differences. Let a_n where $n = 0, 1, ...,$ be a sequence of numbers.

The first differences are:

$$\Delta a_n = a_{n+1} - a_n, \tag{1}$$

the second ones are:

$$\Delta^2 a_n = \Delta a_{n+1} - \Delta a_n, \tag{2}$$

and in general:

$$\Delta^m a_n = \Delta^{m-1} a_{n+1} - \Delta^{m-1} a_n. \tag{3}$$

This leads to the RR:

$$\Delta^m a_n = \sum_{i=0}^{m} (-i)^i \begin{pmatrix} m \\ i \end{pmatrix} a_{m+n-1} \tag{4}$$

From a computing point of view this method only gives an efficient algorithm when the m-th differences are equal to zero. In practice we select the value $m = 6$. This enables to find some simple RR. When the differences are given by a constant value different from 0 the pattern matching identification of this value makes the method impractical.
The other classical methods to identify RR of numbers such as factorization of the a_i, comparison with a known series or the study of the ratio of two successive elements always require an insight knowledge which is hardly implementable in an algoritm.

The following RR are systematically studied:

$$a_n = C(n) a_{n-1} + B(n) a_{n-2} \tag{5}$$

where

$$\begin{matrix} C(n) = c_2 n^2 + c_1 n + c_0 \\ B(n) = b_2 n^2 + b_1 n + b_0 \end{matrix} \tag{6}$$

Equation (5) covers a large set of RR displayed in [9]. The first step consists in solving the system of linear equations given by (5). This gives the possible values for B(n) and C(n) . Then the values for the c's and b's are obtained by solving the system of equations (6) when C(n) and B(n) are not constants.

RR of the following patterns are also considered.

$$a_n = b a_{n-1}^3 + c a_{n-1}^2 + d a_{n-1} + e \tag{7}$$

$$a_n = b' + \sum_{j=1}^{l} c_i a_{n-j} \tag{8}$$

$$a_n = ba_{n-1}a_{n-2} + ca_{n-1} + da_{n-2} + e \tag{9}$$

where b, b', c, d, e, c_i are constants (positive, negative or zero) and l is either given in the input or set to 5 by default. The classes of RR for which b' depends on n (2^n for instance) are not included at present.

Recurring formula generalizing (5), that is of the form

$$a_n = C(n)a_{n-1} + B(n)a_{n-j} \tag{10}$$

where $i, j \neq 1, 2$ are not systematically investigated. To take into account the "Rencontre numbers" we also introduce:

$$a_n = na_{n-1} + (-1)^n \tag{11}$$

This class is considered only when the 10-th term in the sequence is approximately 10 times the 9-th. This property holds also for:

$$a_{n+1} = na_n + b \tag{12}$$

with b small. Many RR consist of a pair of second order recurrences . We check only for the following ones:

$$\begin{cases} a_{2n} = b \ a_n + c \ a_{n-1} \\ a_{2n+1} = d \ a_{n+1} + e \ a_n \end{cases} \tag{13}$$

where the coefficients do not depend on n. It must be noted that according to Jarden [10] a pair of recurrences such as

$$\begin{cases} a_{2n} = c \ a_{2n-1} + b \ a_{2n-2} \\ a_{2n+1} = d \ a_{2n} + e \ a_{2n-1} \end{cases} \tag{14}$$

also obeys:

$$a_{n+1} = (b + cd + e)a_{n-1} - bea_{n-3} \tag{15}$$

which is already handled by the previous RR.

All the sequences of numbers exhibited in [9] which originate from a RR can be derived from those listed above. In fact, many more are thus handled. As stated previously it is not possible to make an exhaustive search for RR. The cost would be prohibitive. Some capabilities to check a RR provided by the user are under study.

At this stage we do not look into recurrent congruential sequences. This in fact only requires to

change the classical arithmetic into modular arithmetic. Algorithms to perform the latter one have been designed for another piece of work and are thus available.
The well known sequences such as Fibonacci and Pell numbers for instance are obviously covered in this approach.
So far we have not been concerned with possible ambiguities in identifying RR. This remark will remain valid for the remaining objects considered in this paper.

Besides synthesizing RR, another obvious facility is provided: to compute a sequence of numbers from a RR. A feature of computer algebra systems is helpful in this work, namely the capability to handle very large numbers.
This purely algebraic approach is not the only posssible one. Since most of RR have been catalogued already, it is conceivable to introduce a database of known RR with their associated sequences of numbers and to search for them. In fact both techniques are pursued.

2.2 Orthogonal Polynomials

Originally, orthogonal polynomials have been studied by Legendre, Laplace, Jacobi and mainly by Chebyshev. Although they can be defined in terms of hypergeometric functions it is more convenient to consider them as a class of their own. The best introduction to this topic is probably the book by Szegö [11]. Besides the classical orthogonal polynomials many classes of associated (to a weight function) orthonormal polynomials have been defined: Krawtchouk, Stilljes-Wigert, Poisson-Charlier, Pollaczek, Meixner, Akhiezer to name a few.
This part of our work does not take into account the analytical properties of these classes of polynomials. Therefore, no special attention is given to polynomials orthogonal on an arbitrary curve (unit circle or Jordan curve) or to the expansion of a given function into a series of orthogonal polynomials [12].

The basic property that we need here is the RR:

$$P_n(x) = (xA_n + B_n)P_{n-1}(x) - C_nP_{n-2}(x) \tag{16}$$

with $n > 1$ and A_n, B_n, C_n constants. It is well known that this is a necessary and sufficient condition for the polynomials P_n to be orthogonal with respect to a given linear functional on a vector space of polynomials (Shohat-Favard theorem). In fact, for our purposes it is more convenient to use the following RR:

$$a_nP_{n+1}(x) = (b_n + xc_n)P_n(x) - d_nP_{n-1}(x) \tag{17}$$

with $n \geq 0$, $P_{-1} = 0$ and $P_1 = 0$.

Having in mind a user oriented approach, we scanned the literature [11, 12, 13, 14 ...] to make a list of the different classes of polynomials which are often found. They appear to be the following ones: Hermite (H_n and H_{ln}), Legendre or spherical (P_n), shifted Legendre (P_n^*), Legendre generalized (P_n^m), Laguerre (L_n), Laguerre generalized ($L_n^{(\alpha)}$), Chebyshev of the first (T_n and S_n) and second (V_n and C_n) kinds, Chebyshev (t_n), shifted Chebyshev of the first (T_n^*) and second (V_n^*) kinds, Jacobi or hypergeometric ($P_n^{(\alpha,\beta)}$) and Jacobi ($G_n(p,q;)$), Gegenbauer or ultraspherical ($C_n^{(\alpha)}$ also often denoted $P_n^{(\alpha)}$), Pollaczek ($P_n(;a,b)$), Charlier ($c_n(;a)$). It should be noted that Jacobi ($Q_n^{(\alpha,\beta)}$) and Legendre (Q_n) functions - not polynomials- of the

Name	a_n	b_n	c_n	d_n
H_n	1	0	2	2n
H_{ln}	1	0	1	n
S_n, C_n	1	0	1	1
T_n, U_n	1	0	2	1
T_n^*, U_n^*	1	-2	4	1
P_n	n+1	0	2n+1	n
P_n^*	n+1	-2n-1	4n+2	n
P_n^m	n-m+1	0	2n+1	n+m
L_n	n+1	2n+1	-1	n
$L_n^{(\alpha)}$	n+1	2n+$\alpha+1$	-1	n+α
$C_n^{(\alpha)}$	n+1	0	2(n+α)	n+2$\alpha - 1$
$P_n(;a,b)$	n+1	2b	2n+1+2a	n
$C_n(;a)$	a	n+a	-1	n
t_n $n = 0,..,N$	n+1	(2n+1)(1-N)	2(2n+1)	n
$P_n^{(\alpha,\beta)}$	$2(n+1).(n$ $+\alpha+\beta+1)$ $.(2n+\alpha+\beta)$	$(2n+\alpha+\beta$ $+1).(\alpha^2-\beta^2)$	$(2n+\alpha+\beta)$ $.(2n+\alpha$ $+\beta+1)$ $.(2n+\alpha+$ $\beta+2)$	$2(n+\alpha)$ $.(n+\beta)$ $.(2n+\alpha$ $+\beta+2)$
$G_n(p,q;)$	$(2n+p-2)$ $.(2n+p-1)$ $.(2n+p)$ $.(2n+p+1)$	$-[2n(n+p)+$ $q(p-1)]$ $.(2n+p-2)$ $.(2n+p)$ $.(2n+p-1)$	$(2n+p-2)$ $.(2n+p-1)$	$n(n+q-1)$ $.(n+p-1)$ $.(n+p+q)$ $.(2n+p+1)$

Table 1: Orthogonal Polynomials

second kind satisfy the same RR as P_n and $P_n^{(\alpha,\beta)}$. Also, the basic RR which links the Jacobi's polynomials is a special case of relations between contiguous Riemann P-functions [15] that we do not consider here.

Table 1 displays the coefficients for the RR shown in equation (17) for the different orthogonal polynomials considered. Our procedure is based upon it. It is trivial to check if some given set of polynomials obeys one of these RR. This is achieved simply by trying all cases for which the coefficients do not depend on a parameter. n is determined by the corresponding location in the input sequence. When one or two parameters must be considered the obvious answer is to solve a system of linear equations to determine the possible solutions and then to check if any holds. The only case for which no strategy has been selected yet is the Chebyshev's t_n [13, vol. 2].

It is possible to find in the literature some more exotic kinds of orthogonal polynomials. For instance, in [16] orthogonal polynomials on a Cantor set are shown with their recurring formula. They will be implemented in our program. Recent advances on setting up constructive methods for non-classical orthogonal polynomials have been achieved by Gautschi [17]. They rely on analytical methods and may be the first really efficient and successful approach to tackle purely analytical problems with computer algebra systems.

2.3 Polynomials

Once again we have to restrict the classes of RR we want to handle. The only general ones upon consideration are second order RR of the general form:

$$A(n,x)P_{n+1}(x) = B(n,x)P_n(x) + C(n,x)P_{n-1}(x) \tag{18}$$

where A, B, and C are at most of second order in n and of first order in x. Their possible values are computed again by solving systems of linear equations. These RR are investigated only upon request of the user in a menu like approach. Indeed, a systematic search yields a prohibitive cost.

Some polynomials have been extensively studied and must be included in such a work. They are the polynomials of Rice, $H_n(\varsigma, p, v)$ [18] and Bateman, $J_n^{u,v}$ and $Z_n(t)$ [19], which may also be written in terms of hypergeometric functions. They satisfy the following RR respectively:

$$n(2n-3)(p+n-1)H_n = \left\{ \begin{array}{l} (2n-1)[(n-2)(p-n+1)+2(n-1)(2n-3) \\ -2(2n-3)(\varsigma+n-1)v]H_{n-1} - (2n-3)[2(n-1)^2 - n(p-n+1) \\ +2(2n-1)(\varsigma-n+1)v]H_{n-2} \\ -(n-2)(2n-1)(p-n+1)H_{n-3} \end{array} \right. \tag{19}$$

$$n(n+u)J_n^{u,v} = \left\{ \begin{array}{l} [(n+v+u/2)(n+u)+(n-1)(2n+v+3u/2-1)-x^2]J_{n-1}^{u,v} \\ -(n-v+u/2-1)(3n+v+3u/2-3)J_{n-2}^{u,v} \\ +(n+v+u/2-1)(n+v+u/2-2)J_{n-3}^{u,v} \end{array} \right. \tag{20}$$

$$n^2(2n-3)Z_n(t) = \left\{ \begin{array}{l} (2n-1)(3n^2-6n+2-2(2n-3)t)Z_{n-1}(t) \\ +(2n-3)(3n^2-6n+2+2(2n-1)t)Z_{n-2}(t) \\ +(2n-1)(n-2)^2 Z_{n-3}(t) \end{array} \right. \tag{21}$$

We do not consider the generalized Bessel polynomials.

2.4 Bessel's Function

Although the relationships for this class of functions are straightforward we have been unable yet to handle them. The main reason is the lack of adequate facilities of the systems at hand to manipulate in a satisfactory way trigonometric objects or Gamma functions for instance. Once an efficient trigonometric package will be available, the handling of these functions will be rather easy. Indeed, the general pattern of RR linking these objects is:

$$B_{\nu-1}(z) + aB_{\nu+1}(z) = (c/z)B_\nu(z) \tag{22}$$

Table 2 shows the coefficients a and b in this RR for the following functions: Bessel of the first kind $J_\nu(z)$, of the 2nd kind (Neumann functions) $N_\nu(z)$, Hankel $H_\nu(z)$, hyperpolic (or modified) Bessel's $I_\nu(z)$ and also $K_\nu(z)$ and the spherical Bessel's $j_\nu(z)$, $n_\nu(z)$, $h_\nu^{(1)}(z)$ and $h_\nu^{(2)}(z)$.

Name	a	b
J_ν,N_ν, H_ν	1	2ν
I_ν	-1	2ν
K_ν	-1	-2ν
j_ν, n_ν, $h_\nu^{(1)}$, $h_\nu^{(2)}$	1	$2\nu + 1$

Table 2: Bessel's and related functions

Anger, Weber, Lommel and Struve functions also satisfy quite simple RR although they include sin, cos or Gamma functions. The incomplete Gamma function exhibits an exponential in its RR. Again, for the reason stated previously they are not handled.

2.5 Hypergeometric functions

For these objects an approach similar to those set up in the previous sections is almost useless. A first reason is that a representation for the hypergeometric function ${}_pF_q(a, b; c; z)$ [20] has to be implemented. If we select a very formal one, such as the one above, this implies that the objects in the input have already been identified and that the problem is solved. A second reason is the number of possible recurrences linking these objects. Even if we restrict ourselves to the confluent hypergeometric function $M(a, b, z)$ [21] and to the Gauss hypergeometric series $F(a, b; c; z)$ (p=2 and q=1 in the general definition) the set of RR is very large. Gauss derived all the RR linking $F(a, b; c; z)$ and its contiguous functions $F(a \pm 1, b \pm 1; c \pm 1; z)$. If one takes also into account other RR listed in the literature [20, 21] 256 of them could be displayed.

Presently the only patterns attempted at are:

$$(b - a)M(a - 1, b, z) + (2a - b + z)M(a, b, z) - aM(a + 1, b, z) = 0 \tag{23}$$

$$b(b - 1)M(a, b - 1, z) + b(1 - b - z)M(a, b, z) + z(b - a)M(a, b + 1, z) = 0 \tag{24}$$

$$(c - a)F(a - 1, b; c; z) + (2a - c - az + bz)F(a, b; c; z) + a(z - 1)F(a + 1, b; c; z) = 0 \tag{25}$$

They are implemented mainly for completeness sake because we do not expect this approach to be really efficient.

A much better method is based on an algorithm coming from the work of Fasenmyer [22]. It is mainly useful for non-orthogonal polynomials expressed in terms of hypergeometric series with $p > 1$ and $q > 0$. It has already been used by Verbaeten [6, 23]. For instance the RR for the bateman's $Z_n(t)$ was derived in [22] using this method which is based on the following remark. When the input sequence is a set of power series $P_i(t)$, calculate $P_n(t)$, $P_{n-i}(t)$ and $tP_{n-i}(t)$ where $i = 1, 2, \ldots$, isolate a common term within the summation symbol (hopefully the one in P_n) and write all these terms with a least common denominator. When neither it nor any numerator in each term exceed a degree n then a linear combination of these terms with (n+1) unknowns must hold (i.e. a RR) and it is straightforward to calculate these unknown coefficients. Such an algorithm is easely implementable. In the example given in [22] only the case $n = 4$ is considered. It is a conjecture that the algorithm is still valid when $n \neq 4$. Some work in this direction has been done already by Verbaeten [23].

Also investigated are the use of term rewriting systems to get a canonical form to represent hypergeometric functions. This is used to formulate the expressions in the input in canonical form. This method is very "heavy" and presents some theoretical problems connected with the completion of the procedure. We need full associativity for instance.
Another possible approach is based on the work of Gosper [3] where some kind of canonical representation seems to be present. Study is under way.

3 Solution of Recurrence Relations

Looking at the literature shows that extensive studies on algorithms to solve RR are in progress [24, 25] and that mainly numerical methods have been developed [26]. The approach we have selected at first is to look at the methods which are taught to first year students in a Computer Science curriculum in a discrete mathematics course.
They are usually the substitution (it is so crude that we disregard it), the charasteristic roots and the generating functions methods. It has been decided to concentrate first on the latter one. Indeed, the charasteristic roots method is rather easy to use, but it is less general than the generating functions one and also it requires to get the exact roots of polynomials. This, in the framework of computer algebra systems is not always implemented.
The generating functions approach relies on an efficient manipulation of power series. Although some implementations of this package are known to have bugs, very efficient algorithms are available. Because of the preceding remark we are first designing a module to handle power series. The different steps involved in solving RR by means of generating functions are described in several textbooks [27 for instance] and do not present any special difficulty. This method is well adapted to an implementation in computer algebra systems. An implementation already does exist in MACSYMA [4]. But, we think that we have a slightly better method at hand and we aim to extend this work.

4 Conclusion

Most of the points we wanted to stress have been covered in the previous sections. We want here to mention a few points related to works in progress in other domains of Computer Science and which are closely connected to RR.
In VLSI circuitry study, systolic arrays (networks of processors which rythmically compute and pass data through a system) require the design of algorithms to evaluate recurrences of the form [28]:

$$x^i = a\, x^{i-1} + b\, x^{i-2} \tag{26}$$

It seems hardly possible to find a link with our approach: we are still far away from programming at the architectural level. But, there is a point in this method which is of direct interest for our purposes: the design of parallel algorithms. Indeed, the advent of parallel processors will have a great impact on this kind of work, although probably not in the near future.

Program transform requires to identify recurrences. This technique, which is used both in Artificial Intelligence and language study, is at the origin of many powerful algorithms detecting recurrences in programs [29 and references therein]. The basic idea is to look for RR by repetitive use of some filtering programs until there is either failure or completion. These algorithms are implemented in Lisp usually. Since we have in mind to introduce some techniques of IA in computer algebra [1] it appears whorthwhile to investigate further the possible links between these different approaches.

A final remark is that implementing trivial mathematical methods allows in our case to build a rather powerful computational tool.

References

1. **J. Calmet and M. Bergman.** *Some Design Principles of a New System for Algebraic and Symbolic Manipulations.* These proceedings.
2. **E.L. Lafferty.** *Hypergeometric Function Reduction - An Adventure in Pattern Matching.* Proc. 1979 MACSYMA User's Conference. MIT Lab. Pub., 465-481.
3. **R.W. Gosper Jr..** *Computer-assisted Strip Mining in Abandoned Ore Fields of the 19-th Century Mathematics.* Talk at the *Conference on Computer Algebra as a Tol for Research in Mathematics and Physics.* New York, April 5-6, 1984.
4. **J. Ivie.** *Some MACSYMA Programs for Solving Difference Equations.* Proc. 1977 MACSYMA User's Conference. MIT Lab Pub.
5. **R.W. Gosper Jr..** *Indefinite Hypergeometric Sums.* Proc. 1977 MACSYMA User's Conference. MIT Lab Pub.
6. **P. Verbaeten.** *The Automatic Construction of Pure Recurrence Relations.* ACM-SIGSAM Bulletin **8**, 96-98, 1974.
7. **J. Calmet and I Cohen.** *Symbolic Manipulation of RR: an Approach to the Manipulation of Special Functions.* In *The Second RIKEN Int. Symp. on Symbolic and Algebraic Computation by Computers.* Ed. N. Inada and T. Soma. World Scientific, 55-65, 1985.
8. **C. Brezinski.** *Accélération de la convergence en Analyse Numérique.* Lecture Notes in Math. **584**, Springer-Verlag, 1977.
9. **N.J.A Sloane.** *A Handbook of Integer Sequences.* Academic Press, N.Y., 1973.
10. **D. Jarden.** *Recurring Sequences.* Riveon Lematematika, Jerusalem, 1966.
11. **G. Szegö .** *Orthogonal Polynomials.* AMS Colloquium Pub. **23**, 1939.
12. **L.Ya. Geronimus.** *Orthogonal Polynomials.* Consultant Bureau, N.Y., 1966.
13. **A. Erdélyi et al..** *Higher Transcendental Functions.* Bateman Manuscript Project, vol. I, II. McGraw-Hill, N.Y., 1953.
14. **R.A. Askey.** *Orthogonal Polynomials and Special Functions.* Proc. Regional Conf. on Applied Math., SIAM, 1975.
15. **E.T. Whittaker and G.N. Watson.** *A Course in Modern Analysis.* 4th ed., Cambridge Univ. Press, 1952.
16. **D. Bessis et al..** *Orthogonal Polynomial on a Family ....* Lett. Math. Phys. **6**, 123-140, 1982.
17. **W. Gautschi.** Talk at the *Journal CAM Conference.* Leuven, July 1984. Proceedings to appear.
18. **S.O. Rice.** *Some properties of* ${}_3F_2(-n, n+1, \varsigma; 1, p; v)$. Duke Math. J. **6**, 108-119, 1940.
19. **H. Bateman.** *Two Systems of Polynomials for the Solution of Laplace's Integral Equation.* Duke Math. J. **2**, 559-577, 1936.
20. **L.J. Slater.** *Generalized Hypergeometric Functions.* Cambridge Univ. Press, 1966.
21. **H. Buchholz.** *The Confluent Hypergeometric Function.* Springer Tracts in Natural Philosophy **15**, 1969.

22. **Sis. M.C. Fasenmyer.** *A Note on Pure Recurrence Relations.* Am. Math. Monthly **56**, 14-17, 1949.

23. **P. Verbaeten.** *Recurrence Formulae for Linear Hypergeometric Functions.* Rep. TW 35, Katholieke Univ. Leuven, 1977.

24. **A. Petrossi and R.M. Burstall.** *Deriving Efficient Algorithms ....* Acta Informatica **18**, 181-206, 1982.

25. **G.S. Lueker.** *Some Techniques for Solving RR.* ACM Comp. Surv. **12**, 419-436, 1980.

26. **W. Gautschi.** *Computational Methods in Special Functions - A Survey.* In *Theory and Applications of Special Functions.* Ed. R.A. Askey. Academic Press. 1-98, 1975.

27. **C.L. Liu.** *Introduction to Combinatorial Mathematics.* McGraw-Hil, N.Y., 1968.

28. **H.T. Kung.** *The Structure of Parallel Algorithms.* Advances in Computer **19**, 65-112, 1980.

29. **E. Papon.** *Algorithmes de détection de relations de récurrences ... .* Thèse de 3ième Cycle. Univ. Paris-Sud Orsay, 1981.

Note: Shortly after this conference was held, a very interesting PhD thesis in Mathematics has appeared. It includes a catalogue of RR and their solutions (**P. Souriac**, *Récurrences avec solutions*, Univ. Sabatier Toulouse, France. Nov. 1984).

Some Design Principles for a Mathematical Knowledge Representation System: A New Approach to Scientific Calculation.

Jacques Calmet (LIFIA, Grenoble, France)

and

Marc Bergman (IIRIAM, Marseille, France)

I- INTRODUCTION

Since its origin, Computer Algebra (CA) has been mainly interested in producing efficient tools to perform algebraic computations. This has been achieved through research in three domains: algebraic algorithms, system development, applications. It must be emphasized that closed links between users and designers has been always a feature of this field. Another line of research deals with architecture of computers for CA. It is pursued recently only.

The first line of resarch is concerned with the design, analysis and implementation of algebraic algorithms. Many achievements have been witnessed in this direction. Many examples can be found in {1}. The second line of research deals with the design and development of Computer Algebra Systems (CAS) {1}. More than 60 CAS have been written over the years. Most of them were built to solve specific problems. Few have general capabilities. Among them are: MACSYMA, REDUCE, SMP, MAPLE. Another important one is ALDES/SAC2 because of its library of well designed algebraic algorithms. Except for the new version of SCRATCHPAD, still under development, all these CAS have a very crude structure when examined from either programming language or software engineering point of views. This structure can be sketched as a three layers one: (1) host language for implementation, (2) programming language, (3) algebraic algorithms and applicative modules.

Because this talk is mainly concerned with system aspects, we do not elaborate on the third layer. It is enough to mention that most of the work in recent years was aiming at producing more and better algorithms implemented through as compact as possible codes.

The selected host language was quite often some dialect of Lisp. This is reflected by checking that the goal of most of the research activity in this domain was to define the right subset of Lisp (or designing new Lisp dialects) in order to improve the portability of CAS. Minimal work has been performed on the programming language layer. In general, a CAS has a programming language in the sense that it allows to write programs based on the available algorithms and to translate them into the host language. But, through any criteria taken from modern language theory it lacks the features associated to this concept.

The structure of CAS is, in many ways, a consequence of the computer technology available when the earlier ones were designed (recent ones have been mainly attempts to do better what previous ones did, except for SCRATCHPAD indeed). General purpose CAS require large memory as well as storage. This requirement was not fulfilled and they had to be designed with minimal code and simple structure. In our domain there was a longstanding belief: there will be a breakthrough for CA when the adequate computers will be available. Adequate means at least 4 Mb of main memory and 100 Mb for storage. The fact is that they are now on the market.

It must be stressed that despite these technical limitations, CAS are very successful. In many fields of Science they have been the only tool to complete some difficult projects {1}. But let us now examine what the situation is nowadays with respect to the possible impact on the design of systems.

(i) Freedom from earlier computer limitations: it is thus possible to experiment new designs and ideas.

(ii) Progresses have been achieved in many domains of Computer Science and they are implementable.

(iii) The "Fifth Generation Software" project in Japan had a great impact in several countries. As a result investigation on what future software will be was stimulated and supported by several agencies. This includes our field.

(iv) We now have a very large user's community. For instance it is quite easy to have access to MACSYMA (this was almost impossible outside of the US a few years ago). Successful CAS (such as REDUCE) are largely distributed and used.

(v) Because of the previous point, CA is no longer restricted to academic communities: some CAS are becomming commercial products, companies are set up to either sell software or for consulting purposes, major manufacturers are hiring experts to design systems. This point is by no means scientific (and in fact is stirring up many controversies!). It must not be overlooked because it will have many obvious consequences.

A first result of all these -very different- facts is that many groups have started to think and work on the design of a new generation of systems. We are aware of at least 15 such projects. All of them are related to CA but they also have another common feature: their capability will go far beyond those of existing CAS. It is not our intention to survey these projects because only a few of them have been presented officially either at conferences or in reports.

We intent to sketch some features of our own project which is now under development. Because of the limited amount of time allowed for presentation and the many points we want to discuss, we have selected the following presentation. We consider the different aspects, features, capabilities of CAS we want to change, improve or include and mention what solution is available. This presentation is certainly not the most accurate one, but it is the only way to be able to stress all the points that we think to be important. Also to fully understand what our project is, implies some familiarity with concepts in different domains of Computer Science such as algebraic specifications, knowledge representation, apprentice and expert systems to name a few.

II- SOME DESIGN PRINCIPLES

The basic idea of this project is that Computer Algebra must be considered as the starting point for a new approach to scientific calculation. Instead of having different pieces of software to perform algebraic computations, graphics and numerical calculations, it is desirable to have an environment integrating all of these capabilities and more. The decision to have a CA package as kernel is based on the belief that more insight is acquired when performing a calculation by such methods rather than by numerical ones.

Another underlying idea is to incorporate as much available knowledge in computer science as possible.

We now proceed to sketch some of the most important points in this project.

1- Portability

This has always been a crucial feature for CAS in the past. Our answer is based upon the assumption that Unix will **the** standard operating system. This system must thus be supported by Unix. For implementation (host) language we use Prolog and Lisp. The interface between the system and the host language must be kept minimal: just a translator having no task to perform besides this one. The Lisp dialect selected is Franz-Lisp but Le Lisp is still under consideration as a possible alternate. It must be stressed that because of the size and scope of the project future translators into (some) other implementation languages can be considered. In this event a list manipulation package will have to be included.

This approach implies that we do not consider portability to be a crucial point. Instead we are aiming at the portability to an easely available operating system and have in mind personal algebra machines as an ultimate goal.

This first point ought to be included in the next section. It is singled out because it has always been an issue in CA.

2- Software Engineering Aspects

Software engineering studies are a prerequisite for such a project. They can be used to plan the development of the system and, at the same time, cure some of the possible drawbacks of CAS.

2.1 Possible inconsistancies between modules of algorithms. Some examples are: two algorithms with same name, global variable set to different values, "global" statements leading to inconsistancies... The latter can be illustrated by a well-known example: if two different modules include the following statements:

(a) $\forall\ x,\ x - x = 0$

(b) It exists a variable for infinity: ∞

then it is possible to reach the rather surprising result: $\infty - \infty = 0$

2.2 The management of CAS is rather crude in many respects. Some examples are that many algorithms are loaded in the main memory but never called, it is difficult -when not impossible- to isolate subsystems specialized to a given >P=O task or problem, to load modules of special algorithms is sometimes left to users, there is only limited control on the extension of a CAS in terms of compatibility and proof of correctness...

The possible solutions are to manage global variables or assignments and algorithm names through dynamical tables. Cross-referencing must be manageable too. To isolate subsystems, a solution consists in marking and garbage collecting algorithms similarly to usual garbage collection. This is feasible in Lisp.

Software engineering is useful to cure some of the existing defects of CAS, but also to plan, manage and master the growth of the system and its fiability. In particular, it may provide aids for debugging.

Although it could be listed under a different heading, we think worthwhile to mention here the definition of standards for algebraic algorithms. Such standards do exist for graphics and for numerical algorithms but not for algebraic ones.

To our knowledge, no attempt to apply software engineering concepts in CA has been published yet. A detailed description of our approach will be find in {2}.

3- Algebraic Specifications and Programming Language Aspects

Semantic correctness is seldom checked in present CAS (but it is in ALDES for instance). Algebraic specifications and their tools are completely absent. Only new SCRATCHPAD {3} does use abstract data types and genericity. We consider programs and abstract data types to be generic when parametrized by types and functions. Term rewriting systems and completion algorithms are very useful tools in that respect.

We consider all these concepts to be mandatory for our purposes. Before sketching briefly our approach, we must stress a few points. We are well aware that these tools are expensive in computing time. Thus, they will not be present in the calculatory parts of the system. Its structure will be as follows: the algebraic specification package will be independent from the other modules. Its main purpose is to check the correctness of semantics and types. It will provide the computer algebra modules with tables of types with their attributes, rules and properties. A user performing a calculation with types already known to the system will not be concerned by these specifications. If he wishes to introduce a new type, he will have two options. The first one is to run the algebraic specification module for the new type. The second one unables to stay in the computer algebra environment by completing the table for types under his own responsability. This way the overhead for manipulating types is kept minimal. Although there are some slight differences with the SCRATCHPAD approach to implement types, both are very similar at computing time.

Other features of programming language theory must be investigated: parallelism, program environment and integration of logic and functional programming. The latter is approached by crude techniques when using Prolog for the algebraic specification and Lisp for other tasks. A more refined treatment is a medium term goal. We do not elaborate further on these points. We just want to mention that parallelism, which is a longer term goal, can be understood as both a way to speed up calculations and to improve the reliability of results by running in parallel two different sets of algorithms performing the same task.

There are several ways to implement term rewriting systems, abstract data types and semantic definition. We present as simply as possible the approach we have selected. For a comprehensive exposition we refer to {4}.
To insure correct semantics we want to design a CAS through an extensible language with a formal semantics basically defined as constructive mathematics. This idea is by no means new. It was already expressed by Musser, Loos, >P=O Jenks-Trager... Improvements in language theory are often achieved through the use of abstract data types and first order logic. Thus we need a constructive methodology for algebraic specifications in the context of predicate logic. Abstract and operational semantics will be both based on a same formalism: the first order logic. The next step is that Martin-Löf's theory of types {5} may be seen as a programming language. It states that one has to define simultaneously terms and types. This implies that it is not necessary to prove that terms are of a certain type. This is our framework.

The implementation starts from Horn clauses (clausal forms of first order logic) to define types and extension of types. Rules for term construction are a set of axioms working as a rewrite rule system supposed to be confluent and noetherian (i.e. finite termination and canonicity) when these properties hold. For a canonical term rewriting system the value of a term is its canonical form. It is well known that canonicity is not always fulfilled. One may have also extension by definition (for instance for square-roots of integers) and many sorted types. This approach can be interpretated as an universe of types. The methodology is extended to meta-types (functor types): application of a type on another one. This allows generic types with recursive definitions. In this framework a CAS is a universe of types which is the context of any calculation and the implementation may be viewed as creating a library of hierarchical types constructed on a universe of types, closed under the basic type operations (convolution, extension, cartesian product ...).

The programming language selected for this part is Prolog. The manipulator of abstract data types with Knuth-Bendix completion algorithm has been completed {6}. Error treatment is in the process of implementation.

Because all the modules are self-contained it is possible to look at other methods for algebraic specification. In this respect, the LPG approach to genericity {7} is under investigation for application in our domain.

4- Software Tools

One necessary condition to have a system as friendly as possible to its users is to make available software tools. They deal with flexible file handling (for instance dynamical creation, comparison, inclusion, connection by names, concatenation, archive ...), sorting, cross-referencing, KWIC-indexing, editing, windowing, text formatting, input/output standards (typesetting, filtering ...), pattern matching. Most of the answers, because of the reference to Unix, are described in {8}.

We list graphics in this section because of the following reason. Graphic capabilities are useful to display outputs (equations for instance) and to plot. The first application belongs to this section. The latter one does not. Plotting is preferable when integrated with calculational tools. The basic remark is that CAS do not make use of today graphic capabilities. Two options are then possible. The first one consists in integrating the packages provided by manufacturers. The second one is preferable. It implies to start from the adopted standards for graphics {9} and to algorithmically program them. This is, at present, a medium term goal for our project.

5- Artificial Intelligence Aspects

It is necessary to enlarge the capabilities of CAS. This need is illustrated by comparing the capabilities of CAS with the table of contents of Mathematical

Review for instance. To base a system on constructive methods only do not permit to achieve this goal. There are two ways to bypass this limitation. The first one is to introduce heuristics and will be mentioned later on. The second one is to rely on IA techniques and methods. The key words in >P=0
this section are: databases, apprentice and expert systems and theorem proving.

Database is a concept seldom used in present CAS. Exceptions are SMP and MACSYMA where a limited use of it is made. A simple way to understand its usefulness is to examine how most of engineering calculations are performed: they mix computations and table look up. We want to introduce databases to offer large table look up and also for building apprentice and expertise capabilities. In many occurences these databases can be dynamically created. For example, databases for orthogonal polynomials and special functions can be constructed using recurrence relations. In turn this feature allow to erase them when no longer needed. This remark is in fact the motivation for {10}.

Many computations are performed many times. When they are either expensive (calculation of some integrals, factorization of some characteristic polynomials ...) or very specialized (Ricci tensor in General Relativity for example) it is convenient to be able to teach the system to learn the corresponding results. This must be driven by the user and by no means a default option. This capability can be extended to wrong branches in a calculation tree later on. The aim is to make possible the construction of data bases of values, results, formulas and even bibliographic informations.

CAS are not expert systems. Although MACSYMA is sometimes listed as one of them, we believe it is not. Indeed its expertise lies mainly in the integration package which could be replaced by the Rish algorithm from the beginning. In the same domain a truly AI model has been designed using REDUCE {11}. It is possible to be build truly expert systems for specialized modules (High Energy Physics, General Relativity, Celestial Mechanics, special functions, definite integrals ...). This means setting up a strategy to get the requested result, references to published works, identification of the computed object in order to check if results are already known, storing results upon request. Most of these expert subsystems have to be designed with experts of the field they belong to.

We do not enter into technical details about the concepts just mentioned. At this stage we just use standard techniques of AI to achieve our goal (plans ...). Specific models will result from this approach.

The last point to be listed in this section is automatic theorem proving. It is well known that many of the techniques used in that respect are quite similar to some used in CA. It is thus natural to include such a module in the project. It will be independent at the beginnig and integrated later on.

6- User's Interface

We collect under this heading two different aspects: interface with the users and pedagogical capabilities. Indeed, two different kind of information are helpful when using a system: one dealing with the state of a computation and another with the methods used.

When performing a calculation it is important to have access to informations on the expected computing time, the state of the computation (for example: loop or long computing time?), the algorithms used ... For the first topic some work has already been completed by different groups. Informations on algorithms, methods and bibliography can be set up through a tracing mechanism in Lisp. This latter facility must be under the control of users, for selected topics only. It must be stressed that pedagogical capabilities are different from the use of a system for teaching purposes.

A last entry in this section is the use of menus which enable some sort of dialogue with the system. The reason for introducing menus will become meaningful in the next section.

7- Algebraic Algorithm Aspects

We do not intend to list all the algorithms we would like to include. From number theory or coding theory to applications in chemistry, there is no limitation! Neither do we mention simplification (it will be very similar to the REDUCE approach) or the style of representation of algorithms (it will be ALDES like, i.e. à la Knuth). We want rather to make general comments.

Algorithms are organized along two different techniques in present CAS. Either as a library (in SAC2 for example) of accessible algorithms or as a "black box" permitting the interactive execution of higher level statements (SOLVE, INTEGRATE ...). Both aspects must be present. This implies possibility to have two different programming modules. To shorten the description, let us just mention that one will be truly interactive but the other one not. The latter is important for algorithm development and testing.

Basically, only constructive methods are implemented. It may be interesting to remind that this was not true for the earlier CAS. Heuristics must be introduced again. The motivation is to be able to tackle problems which are not purely algebraic. Those in Analysis for instance. This is strongly connected to AI techniques as said previously.

As a rule, there is a unique representation permitted for basic objects. SAC2 may be the only counter-example, although that to switch representation is not always possible. It is known that the complexity of some algorithms (multivariate polynomials for instance) depends on the representation. With the introduction of abstract data types and constructors of domains, it is possible to make progresses in this direction.

Usually only one method to perform a given task is implemented. If two are, (factorization of polynomials for instance) the user has little initiative to >P=O select the one he prefers. For the same example one can list methods by Kronecker, Berlekamp, Cantor-Zassenhaus, Camion-Lazard, Lenstra, Kartofen ..., but they are usually not available. A new methodology will consists in making many methods usable. One is the default option. Others are presented through menus. This is a way not to choose between "best algorithms" and most efficient ones. In the same spirit, we need more statistics on data to be able to answer the following question: when are "crude" algorithms sufficient? Crude means that they may not solve all possible cases but are very straightforward. The possible strategy is to introduce firstly crude algorithms as a first try for success for basic operations (gcd, factorization) and then use statistics and measurements to develop a heuristic to try to get optimal computing times.

8- Software Integration Aspects

We refer here to the integration of graphic, symbolic and numeric packages. The answer for the first two ones has been given in a previous section. Symbolic to numeric interface is a feasible task. Many existing CAS can generate Fortran programs or at least parts of programs. The numeric to symbolic interface is still a research topic. We must add that for our original project this is not a priority goal.

III- CONCLUSION

This is a rather sketchy presentation of a project under development. It must be mentioned that this project may evolve through collaborations with other groups. The consequence would be that many goals listed as medium or long term ones would become short term ones. The main characteristics of the project would not be modified. They can be summarized as follows:

(i) System for scientific computation and mathematical knowledge representation, oriented toward both users and algorithm designers.

(ii) Many modules distributed to specialized tasks with inter-connections among them. Possibility to tailor subsystems for special use.

(iii) Apprentice and expertise capabilities.

(iv) Strong algebraic specifications.

(v) Integration of softwares from different origins.

(vi) Different from present CAS but capable to do what they do and much more.

(vii) Same need for good algebraic algorithms.

The final remark is that several pieces of code are already available for algebraic specification and algebraic algorithms. This enables to have a prototype in just over three years.

REFERENCES

1- "Computer Algebra". Ed.by B. Buchberger et al., 2nd edit. Springer-Verlag, Wien (1983).

2- J. Calmet and D. Lugiez, "Software Engineering Aspects of a Mathematical Knowledge-Based System". LIFIA report (1986).

3- R.D. Jenks, "A Primer: 11 Keys to New Scratchpad". Springer Verlag LNCS **174**, pp 123-147, (1984).

4- M. Bergman, "Algebraic specifications: a constructive methodology in logic programming". Springer Verlag LNCS **72**, pp 91-100, (1982).

5- P. Martin-Löf, "An Intuitionistic Theory of Types". Proc. Logic Coll. 1973, Ed. by H.E. Rose et al., North-Holland, Amsterdam, pp 22-29, (1975).

6- M. Elbaamran, "Etude et réalisation d'un système d'aide à la spécification algébrique des types abstraits". 3rd cycle thesis, Univ. Aix-Marseille (1984)

7- D. Bert, "LPG Manuel", LIFIA-IMAG Report, unpublished (1984).

8- B.W. Kernighan and P.J. Plauger, "Software Tools in Pascal". Addison-Wesley, (1981).

9- Enderle, Karsy and Pfaff, "Computer Graphics Programming, GKS: The Graphic Standard". Springer-Verlag (1982).

10- J. Calmet, "Manipulation of Recurrence Relations in Computer Algebra". This Conference.

11- M. Vivet, "CAMELIA". PhD thesis, Univ. du Maine, unpublished (1984).